An Exercise Workbook

University of Calgary

Long Grove, Illinois

For information about this book, contact:
Waveland Press, Inc.
4180 IL Route 83, Suite 101
Long Grove, IL 60047-9580
(847) 634-0081
info@waveland.com
www.waveland.com

10-digit ISBN 1-57766-165-6
13-digit ISBN 978-1-57766-165-8

Printed in the United States of America

10 9 8 7 6 5

For Sandra
Toughest editor, critic, and
most enthusiastic cheerleader.

In Memory of Richard 1946–2000.

And
for all those who have served
as “guinea pigs”
during development of these exercises.

Contents

Preface

The discipline of primatology, and especially the field of primate behavioral studies, is extremely attractive to students. It promises excitement, strange and fascinating animals, exotic locations, and, perhaps, even public acclaim on occasion. I have wandered along the pathways of the discipline for over thirty years, always fascinated by the behavior, morphology, physiology, but especially the ecology and evolution of our nearest mammalian relatives. Occasionally I even succeed in passing some of that enthusiasm on to students, who go on to their own research careers. Few individual workers achieve the heights of fame and become well-known public figures. Those who do, like Goodall, Galdíkas, and Fossey, become public icons. The majority of us simply "labor in the trenches," teaching and, whenever possible, conducting our research programs.

Alas, many students only see the flash and the public attention. They overlook the hard work that is necessary to aspire to membership in the profession. Many students do come to recognize that primatology offers a life of hard work, sometimes hard physical labor under uncomfortable, even dangerous conditions, and they accept that self-discipline and incremental progression are the cornerstones of success. I have sought to convey the fascination, ecstasy, and frequent disappointment inherent in scientific research to these students. Throughout my teaching career I have tried to place an emphasis upon accuracy, precision, and effective communication of results. (As some editors might be inclined to comment, this latter is a case of "do as I say, not as I do.") These exercises are an outgrowth of those principles and practices.

The first edition of this workbook was the result of more than half a decade of experimentation in the process of teaching the skills and techniques of observation. In the interval since its publication, I have learned—just a few new things—and this second edition is much larger and deals with a number of new concepts. The majority of the text has been updated and, in many areas, completely rewritten. The statistics have been substantially revised and grouped into categories, and the practicalities of calculating a standard deviation have been added. A major expansion involves a suite of field ecology exercises that several colleagues suggested would be an enhancement and make the book useful at field school sites.

The exercises in this series remain as they were originally, based solidly in the scientific method and grounded upon empiricism. I hope that they will bring students into some proximity with primates as subjects of observation, and with the realities of scientific inquiry. It should be recognized that even if a primate colony or zoological collection is not available, the observational exercises can still be carried out on squirrels, ungulates of both domestic and wild forms, on feral dogs and cats (with suitable consideration of the dangers!), and the entire avian realm. All of the original exercises were tested, exploiting several hundred students as "guinea pigs" over several years. Many of the remaining ones have been developed with the aid of senior undergraduate and graduate students but have not been directly tested at the sophomore level.

Anyone who conceives a teaching exercise that may be suitable for the next edition of this workbook is encouraged to communicate with me about it. Messages can be sent via e-mail to paterson@ucalgary.ca at any time.

These exercises owe some of their structure to many colleagues. Among the earliest sources of improvement were Andrew Petto and Claud Bramblett. Their advice and suggestions have become so embedded in the work that I can no longer point to their locations. Other ideas emerged from discussions over the years with Francis Burton, Anthony Coelho Jr., Takamasa Koyama, Pamela Asquith, Mary Pavelka, Sue Taylor Parker, Anne Russon, and Lisa Gould. Several graduate student teaching assistants, who will be grateful to remain anonymous, have also been involved. I must also thank the curators and primate keepers of the Calgary Zoological Park (some of them were my students during the development phase, and continue to cheerfully accept the horde of young primates who invade each fall term). I thank Tom Curtin of Waveland Press for again undertaking the risk of publishing this version. I especially appreciate the efforts of Jeni Ogilvie and Sandy Smith for their detailed editing that corrected a plentitude of my persistent errors.

The employment of the exercises in this volume is naturally up to the individual instructor; however, the pattern in which they are used at Calgary where we run thirteen-week semesters followed by a two-week examination period, may be of some interest. The complete set would be more than needed for a full-year course. I employ a subset, usually four or five exercises, in a semester. These are a different set each time the course is run, but usually two are from the preliminaries, and two, with higher weighting, from the main methodology forms. Graduate students who may lack observational experience or methods training are usually required to conduct all of the exercises, and to perform exercise 7 using a computer logger system to collect thirty hours of data. This pattern of use has led to significant improvement in the observational skills of our students over the years.

THE CD

It is a standard ISO 9660 format CD-ROM, and hence should be readable by any Windows or Macintosh platform computer. The first instruction is to open and examine the Read Me file. This will provide information on the following files contained on the CD: Extras, Forms, Programs, Video, and Worksheets.

J. D. Paterson
Airdrie and Calgary

Part 1

The Study of Behavior

1

An Introduction to the Observation of Nonhuman Primate Behavior

What is "behavior"? In a strict, technical sense, it is the actions carried out by an individual subject, and as such is a real phenomenon. However, in the common use of the term as employed by most people and primatologists, it is the perceived and recorded, "record" of those actions. Since behavior is internalized and interpreted by an observer, it is subject to errors, and, as Gary Larson suggests in the cartoon reproduced on the next page, such "processed" observations may have unforeseen consequences.

Primatologists observe the behavior of nonhuman primates for the purposes of learning about the relationships between their subjects and the following:

- the environment occupied by the species;
- the particular way in which the species exploits its habitat;
- the social organization of primate groups; and
- the anatomy and biomechanics of members of the species.

To the extent that we can answer these questions, we can make better inferences about the behavior and social organization of our ancestors than we can from the study of their fossilized bones, their tool remnants, and their living sites alone.

Please note, however, that there is *no* assumption that any of our remote ancestors were very much like *any* of the extant nonhuman primates. *There is* an assumption that similar problems of adaptation to a particular environment may lead to similar anatomical and behavioral solutions to those problems.

In the exercises presented in this book, students will be asked to observe the social behavior of nonhuman primates, mostly those living in zoos, and to write reports about their observations. However, it is not a requirement of observational training that primates be used. A student could as easily study urban dogs, or cats, squirrels, raccoons—indeed virtually any mammal

THE FAR SIDE by Gary Larson

"Shh. Listen! There's more: 'I've named the male with the big ears Bozo, and he is surely the nerd of the social group—a primate bimbo, if you will.'"

is suitable, though one is cautioned against performing studies on human primates due to legal and ethical complications.

Since the adaptive requirements of surviving in a zoo are clearly different in many respects from those of living in the wild, *many kinds of behavior may be affected; behavioral patterns may be suppressed, enhanced (elevated in frequency or intensity of performance), or distorted*, and this should be remembered in studying the results of observations on captive subjects. A student who observes carefully can learn much about the social behavior of these animals and consequently much about how to observe the behavior of free-living primates, including that of *Homo sapiens*.

Students are often surprised at the close resemblances between some kinds of primate behavior and the behavior of humans in similar situations. (From this point on, for the sake of convenience, the use of "primate" will refer to "nonhuman primate.") These similarities provide evidence that primate behavior, like primate anatomy, is derived *in part* from a common phylogenetic lineage.

However, one must be extremely careful NOT to impute human motives, emotions, or intentions to these animals. This is *anthropomorphism*, and is an often unconscious form of bias that is associated with *anthropocentrism*, the perspective that holds man (*Homo sapiens*) as the central and most important organism in existence. While most anthropologists and primatologists decry the anthropomorphic and anthropocentric perspectives as unacceptable bias, a few researchers have followed the lead of Asquith (1986, 1997) in admitting a degree of anthropomorphism into their study methodology. A number of recent works (collected in an edited volume by Mitchell, Thompson, and Miles, 1997) have pointed out that it is impossible to expunge anthropomorphism from primatological study since it is a sub-category of "primatomorphism," which humans share with other primates. In spite of this current acceptance of some anthropomorphic influence in primatology, it ought to be clear after a few minutes thought that these primates are NOT little humans, and that even IF the action does reflect the same emotion, motive, or intention as a human might express under the same circumstances, THE OBSERVER CANNOT PROVE IT. This is an important principle to keep in mind throughout all of the observational exercises and in all scientific enterprise.

A Bit of History

As Loy and Peters (1991) have noted, the observation of other animals is an ancient avocation. Aristotle advocated it, and Darwin actualized such studies with his *The Expression of the Emotions in Man and Animals* in 1872. Field observations of monkeys and apes have been conducted on an intensive and continuing scale by North Americans and Europeans only since 1958–59, although a number of noteworthy observational works were conducted during the 1930s and 1940s: fieldwork most notably by Americans Clarence Ray Carpenter and Sherwood Washburn, the Japanese studies of macaques beginning in 1948, and zoo studies by Sir Solly Zuckerman (1932). Other work, particularly in psychology, on the intelligence and learning capabilities of apes, goes back to the experiments conducted by Wolfgang Kohler at the Canary Island Laboratory during the First World War. The period since 1958–59 has been characterized by a continual progression in the skill of researchers and the development of new techniques in field research. Particularly thorough observations have been carried out on a number of monkey species—especially the baboons (*Papio cynocephalus*, although the taxonomic community now recognizes *Papio hamadryas* as the correct name), the members of the macaque monkey group (*Macaca*), howler monkeys (*Alouatta*), and on the larger apes—East African chimpanzees (*Pan troglodytes*), and the mountain gorilla (*Gorilla gorilla beringei*). These studies have resulted in a wealth of new information, which has tended to alter our ideas about the behavior of primates, in sometimes quite unexpected ways.

Scientists of many disciplines have converged on the study of primate behavior: *anthropologists* seeking the nature of man and models for the development of human society; *psychologists* seeking the nature of thought processes, and the basics of learning; *linguists* seeking to understand the origins of language, and testing concepts of grammar; and *zoologists* challenged by the most complex of nonhuman animals. As these disciplines converge, with their traditional attitudes and techniques, interests begin to blend, new methods and new techniques emerge, resulting in even more productive field observations. This blending of concept and technique has also enabled us to talk more realistically about functions and behavior in extinct primates, including our own more immediate ancestors.

The Study of Behavior

Behavior is *what* an animal does. That behavior is an exceedingly complicated issue and the study of behavior is similar in complexity. A succinct statement on this is the familiar quotation from Alexander:

> The study of behavior encompasses all of the movements and sensations by which animals and men mediate their relationships with their external environments—physical, biotic, and social. No scientific field is more complex, and none is more central to human problems and aspirations. (Alexander, 1975: 77)

Since it is our purpose to develop some skill in the study of behavior, we all need to be aware of what it is we are studying and the how and why of doing so. All ethologists (those who study behavior) and most primatologists (those who study primates) will recognize the four areas of behavioral study formalized by Niko Tinbergen (1963) and often phrased as questions:

1. What is the *function* of the behavior? In other terms what does the behavior do for the animal, and what are the consequences of performing it? It can be thought of as being orientated toward understanding the adaptive value or survival value of the action.
2. What is the *causation* behind the behavior? What mechanisms or conditions lead to the presentation of the behavior? This question can lead into the complexities of physiology and learning as well as morphology and ecology.
3. What is the *ontogeny* of the behavior? The focus turns to how the behavior pattern may be developed in an individual, and may extend into attempts to understand how the patterns and processes of life history may modify the behavior over time.
4. How did the behavior *evolve* in the species displaying it? This central theme of modern biology leads to considerations of phylogenetic, genetic, and "cultural" patterns of distribution within species and across larger groupings. Recent examples such as Sillen-Tullberg and Møller (1993), Di Fiore and Rendall (1994), and Rendall and Di Fiore (1995) deal with aspects of the evolution of primate sexual patterns.

An alternative view of these four questions is that they are four *aims* of ethological study. Indeed they serve well in both contextual frameworks, but it is in this latter frame that Burghardt (1997) has argued that a *fifth aim* can and should be added to ethology. This fifth aim could be phrased as "What is the private experience of the animal presenting the behavior?" Burghardt suggests that alternate identifiers for this concept would be "personal world, descriptive mentalism, subjective experience, and heterophenomenology," and the objective is to "identify patterns and processes in life as experienced." This might generally be interpreted as the life history of an individual. This aim is well beyond the intentions and orientations of this manual.

Tinbergen's four questions—function, causation, ontogeny, and evolution—should be the base stratum upon which all primate behavioral observation is founded. The four can be compressed into two sets variously contrasted as "proximal" and "ultimate" causation (Wilson, 1975) or "how" and "why" (Alcock, 1989), and thus serve as immediate guides toward an understanding of the behaviors being examined. In most cases, "proximal" or "how" questions are addressed to the level of individuals and individual behavior, while "ultimate" or "why" questions, being evolutionary, must be addressed to groups, populations, and species.

The Observer: Perception, Errors, and Observer Effects

The engine that drives observational study is the human observer, and any discipline that relies upon human perceptions for its primary data gathering has a problem. Indeed, not just a single problem, but a complex and interacting set of them is normally present. As we shall see later in chapter 2, many forms of bias exist for the observer, depending upon his or her sex, age, and cultural background. But before these issues there are others, ones strongly based in the biological structure of the observer.

One assumption often imbedded in us is that we "see" with our eyes. Sadly to say, the human optical system is less than perfect, often very poor, and the direct physical information of photons striking particular retinal cells is modified, smoothed, improved, and interpreted before it manifests itself in the visual centers located in the posterior aspect of the brain, as part of an image in our mind. The physical deficiencies of the optical system, its blind spots, and the neural pathways and mechanisms of vision are well beyond the intent and needs of this book, however, the fact that these exist should make any observer cautious about strongly affirming "This is what I saw!" Perception varies between individuals and in modes of operation. My wife is very proficient at locating animals in the forest, far more so than I am, but I am more proficient at discerning fine scale movements by individuals once they have been located. Thus we differ in perceptual capabilities but complement each other in the field. The inate differences between observers can be trained toward a common ground, an issue that is taken up in the exercise on Interobserver Reliability (exercise 12).

Since we have so many internal links from incoming photon to realized image, there is opportunity for errors to creep into our perception. One possible phenomenon is that the mind insists upon seeing an arrested action as continuing; on seeing something in the "blind spot" where there are no retinal receptor cells, primarily by stretching the surrounding image to cover the "spot" or by interpreting something as "seen" even though it is actually out of sight or obscured. Add to these problems the possibility that the intellectual operations of the observer may also create an "error of apprehension," that is, seeing the actions incorrectly. Finally, the observer may incorrectly record the observation, the fingers doing one thing while the brain has apparently commanded something else.

And to further confuse the issue, we must note that the observer will also have an interactive influence on the subject. That is, the simple presence of the observer may cause changes in the behavior of the subject animal. This may be an immediate provocation of a flight response or an "uneasy" tolerance of the observer until he or she approaches too closely, whereupon flight takes place. In the early reports of primate field studies, it was common to find notations of the flight distances observed at first contact and the time required for the subjects to acclimate to a minimal observing distance. While these features are no longer commonplace parts of scientific reports, it remains the situation that a substantial amount of time may be required in order to achieve regular observation of wild populations. The observer should not assume that acclimated subjects will tolerate observers at any location or in any numbers. As an example, the well-acclimated Sonso baboon troop that was the subject of a study in 1996 were less than tolerant of three observers (myself, my wife, and our tracker) when inside the forest. On several occasions, with myself in the lead, we would be surrounded and intensively threatened at distances of 1 to 2 meters; in other locations, the troop would not allow us to approach within 30 meters. Yet on other occasions, when working alone, I spent several hours with the entire troop resting and grooming within 5 meters. The take-home lesson from all of this is that the relationship between the observer and the observed is an interactive one, and variable in time and space. This has an influence on the quality and quantity of data that can be collected, and to some extent is responsible for a bias in the data.

Wild, Free-Ranging, and Caged: Similarities and Differences

Behavioral studies of primates can take place in three different conditions: in their own natural environment where they may or may not be constrained and influenced by human populations; in large group enclosures where their activities are constrained by the presence of fences or moats and the interactions with their keepers; and in typical zoo cages, surrounded on all sides by human fabrications, often with some simplified structures to encourage "naturalistic" behavior.

The differences between these three conditions might initially suggest that the behavior patterns of subject primates might be substantially different in each. However, to a great degree, the "form" of species-specific behaviors will be uniform in all environments; there may be minor variations on the patterns, but the main differences will be in the frequency and intensity of the acts. The general tendency in most primate species is for a change in frequency and hence the rate at which specific behaviors are performed, as the observer changes from natural to free-ranging to caged environments. The directionality of the rate change cannot be reliably predicted as there are a large number of "intervening" variables between the three states, but it has often been noted that both aggression and grooming increase under cage conditions.

PRIMATOLOGY AND ETHICS

Throughout the industrialized world, wherever research employs animals as subjects, there have arisen groups who find this activity repugnant and unethical. Many of these groups have resorted to extreme means to assert their position and to attempt to impose their views upon the scientific and lay communities. Fortunately, the observation of behavior alone has generally been found acceptable by these groups, but some do advocate a complete cessation of all animal-related research, including observation. The policing of ethics in relation to scientific research is not a simple problem, but as students of behavior, ethical standards for all research work must be adhered to. This is generally easier for practitioners of observation than for other projects in that there is normally no direct interaction with the observed, and consequently no physical pain will be inflicted. However, manipulation of the care conditions, the cage or enclosure size, feeding schedules, and so forth, can have indirect effects on the subjects and hence require careful ethical consideration before implementation is allowed. There are far fewer ethical problems associated with naturalistic observation in the subjects' native environment, but it remains the case that most, if not all, projects will require ethical approval either by the observer's home institution or by the government of the area where the primates are to be studied.

In North America, every university and research institution has an animal care committee (ACC) of some form whose responsibility it is to oversee the implementation of the institution's ethical guidelines and to certify that all research, at all levels, is done in compliance with these guidelines. This implies that as a student conducting the exercises in this book, you would also be expected to submit your research proposals for certification. However, most ACCs are overloaded with project applications, and, consequently, it is almost certainly the case that your instructor has done this for you, and the course is operating under a blanket approval.

For research conducted in "source" countries where primates reside, there is normally a governmental branch or agency that is charged with

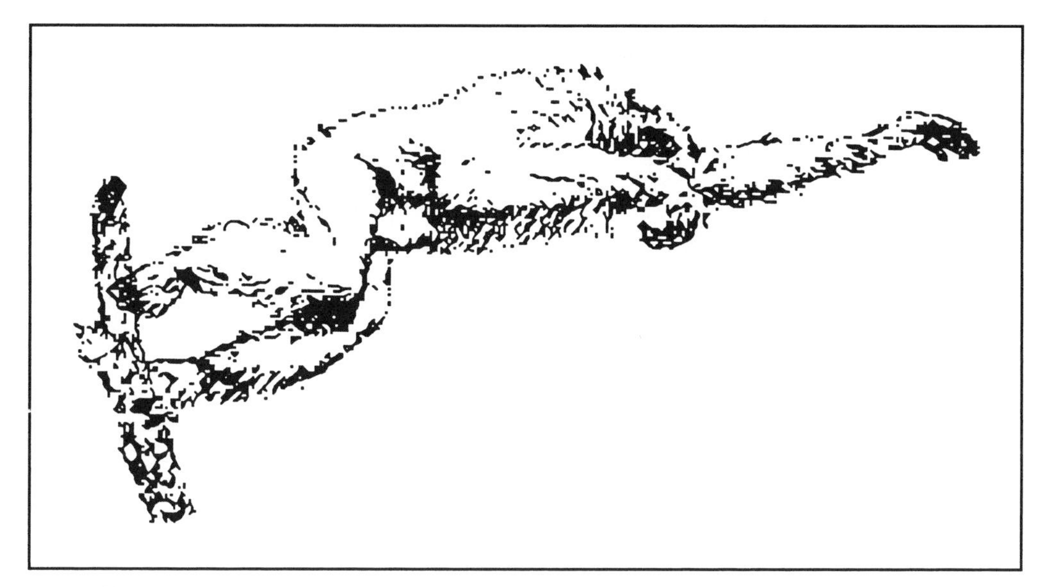

Take-off phase of a *Hylobates lar* leap (after Oxnard 1983:119)

oversight of research projects undertaken within their boundaries. The forms, requirements, lead times, and fees charged are specific to each country. Anyone embarking on a field project needs to take into consideration these costs in time, money, and effort. Contact with researchers who have done work in the target country would be the best source of information on the requirements and protocols. And established field stations, most of which can be located through listings in the *International Directory of Primatology*, can usually ease the process.

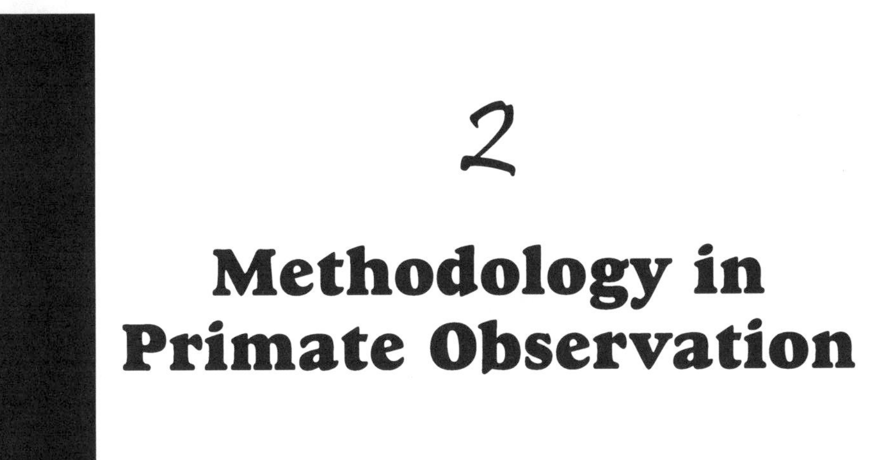

2

Methodology in Primate Observation

THE SCIENTIFIC METHOD

Western civilization has produced two paradigms or operational procedures for learning about the universe around us: scientific and mystic. The mystical paradigm has a number of flaws in spite of its widespread acceptance and use. It is not testable in any way, it relies upon imperceptible and unmeasurable forces, and it is most often "revealed" to the ignorant. The alternative, the scientific paradigm, while not perfect, is the best process yet developed to examine nature. It is a testable system, relies upon perceptible and measurable forces and mechanisms, and has achieved the status of being the only accepted paradigm within the scholarly community.

The current paradigm in science holds that the scientific method is actually a simple process. First, a hypothesis is generated. This is really a guess as to how the situation under study operates or can be explained. Second, predictions of results are generated from the hypothesis. In other words, if our hypothesis is correct then the results from an experiment or study of this form will be x and y. Third, we either conduct an experiment or proceed to collect data. This process is dependent upon careful design of experiments, or careful and precise observation. Fourth, the data is analyzed to determine whether or not it contradicts the results projected from the hypothesis. At this point it is important to note that the entire scientific method of inquiry is directed towards *rejecting* or *invalidating* a hypothesis. One can NEVER prove a hypothesis. Fifth, the findings are reevaluated. If the data contradicts the hypothesis, the original guess is at least partially incorrect (one might have constructed an incorrect or defective hypothesis from the original statement), and two procedures become necessary: replication of the experiment or observations to confirm that an experimental or observational error was not made and revision of the hypothesis—starting over from a new guess. If, on the small chance that the hypothesis and guess are *not* rejected, then we can provisionally accept them, but in doing so, numerous new questions will arise. This entire process indicates that the scientific method is a repetitive cycle. This should not be construed as implying a "circular argument," but a helix of operations, in which an incremental improvement in explanation is achieved with each "circuit" of the process.

The process as applied to the study of behavior can be diagrammed as follows.

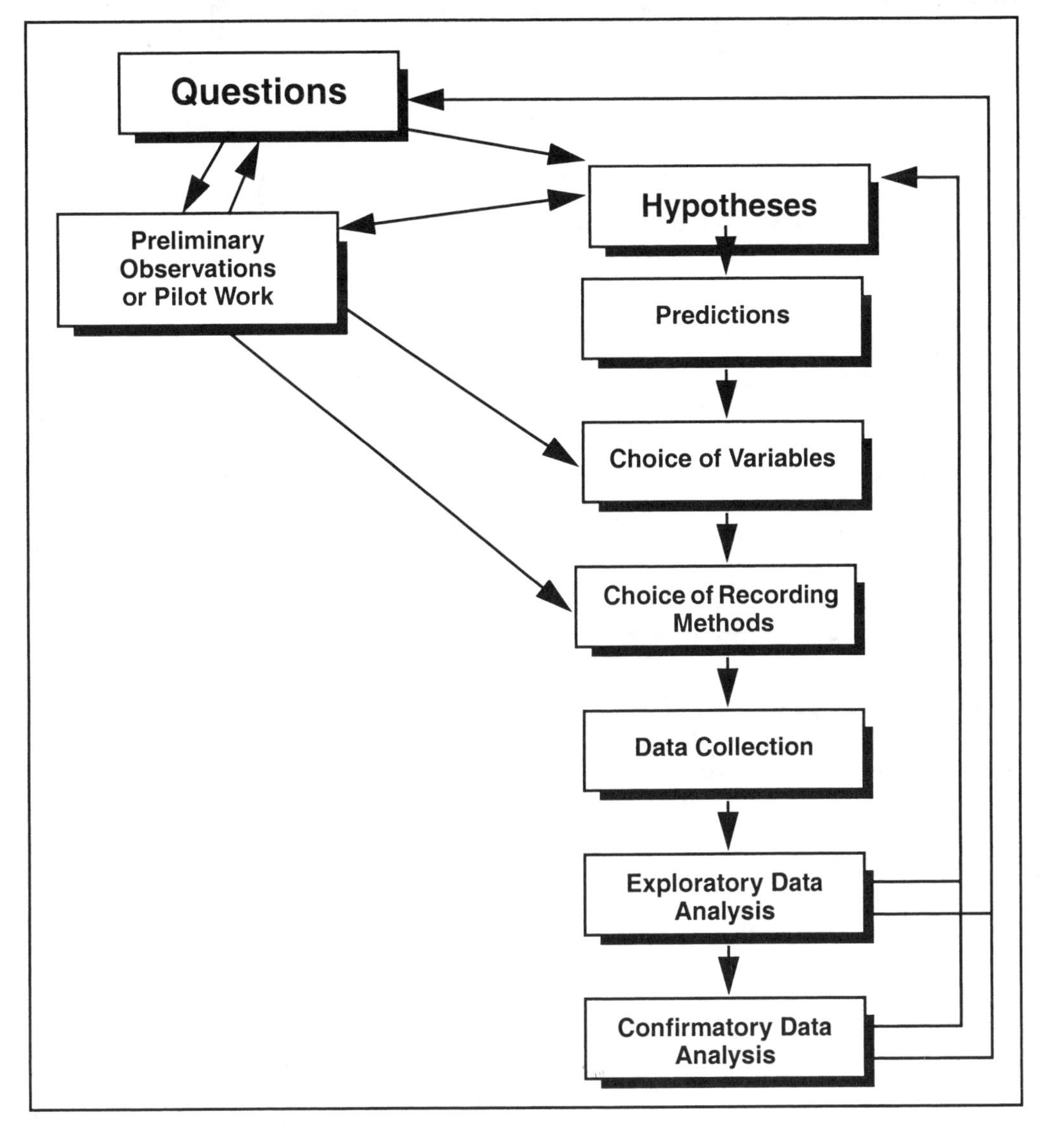

Figure 2.1 The process of studying behavior.

One important aspect of Figure 2.1 is the box representing "Preliminary Observations or Pilot Work." Throughout primatology, and ethology, the costs associated with research are constantly influencing the conduct of research, and it would be very ill advised for a researcher to start with a research question, a set of hypotheses, and then proceed to collect data for a long period, perhaps years, before attempting any analysis. What does the worker do if the results show that the research question or the hypotheses were inappropriate? The safeguard against calamity is the preliminary work or test project, a trial run in which the whole project design is tested, verified, and modified as needed for the testing of the hypothesis or hypotheses. It must be repeated that this process is NOT a correction needed to obtain "correct" results, but is one intended to make certain that the hypotheses and variables are actually doing what the research design intended. Of

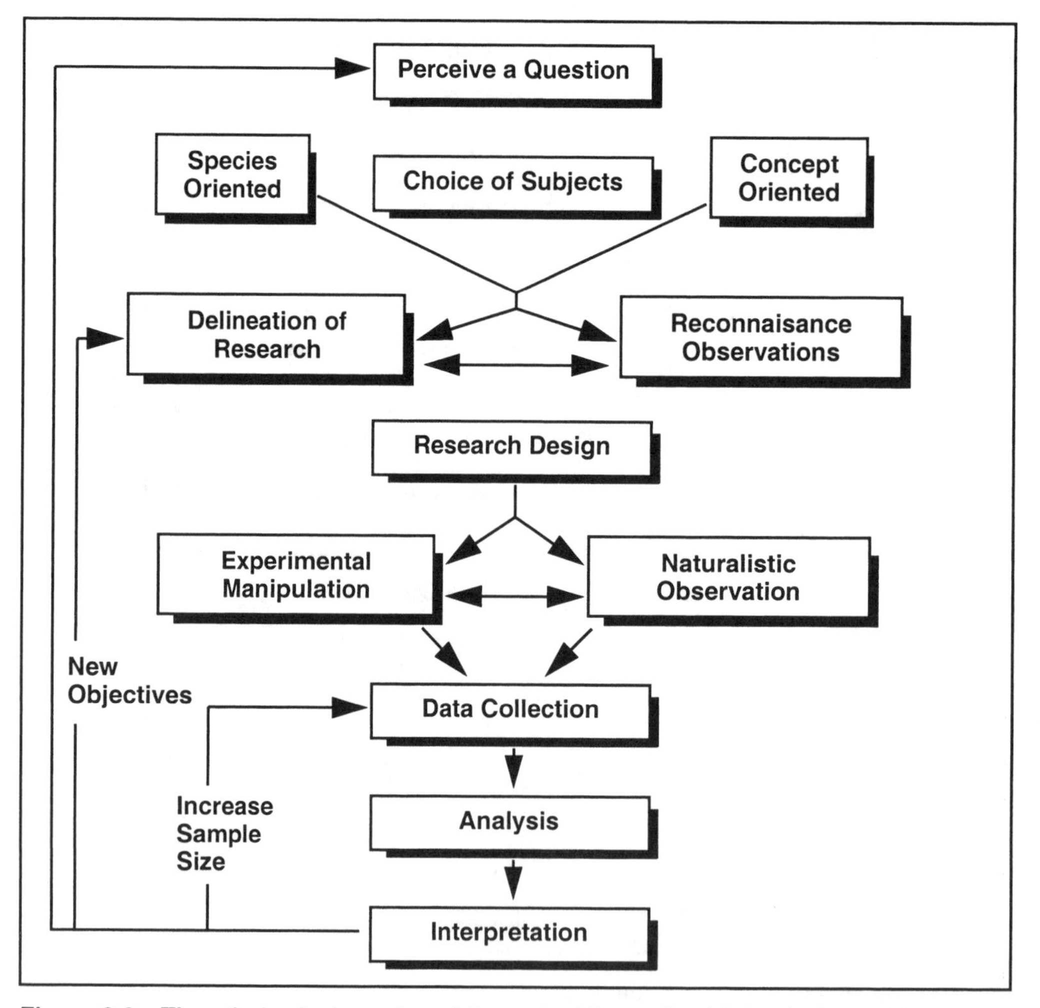

Figure 2.2 The ethological version of the scientific method (after Lehner, 1996:12).

course it is possible for this test to show that the hypotheses are quite incorrect and thus lead the researcher to a major revision of the project. It is always better to have the project tested and verified to some degree before actually embarking on the fieldwork phase of study.

Within the general field of primatology and ethology, the model can be elaborated somewhat to encompass the orientations toward particular species or concepts and the distinctions between experimental and naturalistic approaches (see Figure 2.2 on p. 16). The difference between "experimental" and "naturalistic" implies that the former involves studies of subjects conducted under captive conditions, normally where almost all aspects of primate existence can be controlled, while the latter implies a lack of experimental control and essentially a free and unconstrained observation of the subjects. The discrimination between species-centered research and that which revolves around concepts is quite straightforward. With species-centered research, the orientation is toward finding out everything possible about the subject species, and draws upon all questions and concepts that apply to it. With concept-centered research, where a particular behavior or mechanism or model is under examination, this drives the choice of subject species. In this case the August Krogh Principle (Krebs 1975; Lehner, 1996) comes into play as a guide to selection of an appropriate species to employ for the study (see Lehner, 1996:50–53 for a detailed description). However, this case is less important for most primatologists who are species focused, or, at least, constrain their interests to the boundaries of the Order.

Sampling Methods and Biases

One of the most important revolutions in field primatology methods came with the publication in 1974 of a paper by Jeanne Altmann in the journal *Behaviour*. Dr. Altmann pointed out that *the techniques of taking field notes that were in use up to that time were grossly inadequate for any acceptable statistical analysis* and, furthermore, that most observations were taken without any cognizance of, or correction for, a number of biases. The biases that she pointed out were often obvious but unconsidered sources of error in "statistical" calculations presented in reports. For instance, she noted that male fieldworkers tended to overemphasize the importance of the role of the males in the social matrix of a group, and female workers did the converse. Similarly, there was an "accidental" bias in that highly visible activities were more frequently recorded than were quiet, unobtrusive activities.

Also, it was evident that most workers did not take adequate care to separate two important classes of behavior—those of "states of behavior" and "events of behavior." This distinction is important because the two categories represent distinct *kinds* of data, and they *must* be treated statistically in different ways. A behavioral event is one that is instantaneous, momentary, and of very short duration, or the onset of a state behavior may be treated as an event. That is, in general terms, a behavioral event occurs in a very short period of time, and it is difficult or unsuitable to attempt to measure the duration. A behavioral state, on the other hand, is durational behavior, an activity that displays an appreciable expanse of time that can be measured. These may be formally defined in this way:

event: Instantaneous or momentary behavior that occurs without measurable duration. The onset of any behavior may also be considered as an event.

state: Behavior with a measurable duration (durational behavior), but may refer to any behavior at a given instant in time.

However, the distinctions are somewhat arbitrary since they depend upon the capability of the observer to record the time duration involved. This limitation varies from observer to observer, and especially so when technological aids to observation are employed. It is straightforward to consider the onset of a state behavior as an event, and one can refer to any behavior at a given instant in time as a state. Theoretically, if we could observe and record on a millisecond scale, we could treat all behavior as state behavior since it would be measurable in time. However, the practicalities of observational research even when using a computer data recorder mean that any behavior taking less than one second is recorded as an event.

Consider the following example of a partial observational record.

Record of Observation		**Behavior Form**
10:32:22	Jo-Jo receives grooming from Jay-Jay	state—134 seconds
10:34:36	Jo-Jo sits up & Scans	event
10:34:37	Jo-Jo locomotes to food dish	state—4 seconds
10:34:41	Jo-Jo eating	state—91 seconds
10:36:12	Jo-Jo drinks	event
10:36:12	Jo-Jo scratches	event
10:36:13	Jo-Jo chases Spike	state—42 seconds
10:36:55	Jo-Jo receives threat from Alice	event
10:36:56	Jo-Jo presents to Alice	event
10:36:57	Jo-Jo receives grooming from Alice	state—uncalculated

The right column is *not* recorded at the time of observation, but represents an after-the-fact decision tree record by the observer. This type of process is necessary to divide the behaviors into the categories of event and state. The categorization of a particular behavior depends upon two factors: (1) the actual duration of the behavior—obviously some activities MUST be recorded as events since the duration may be too short to measure accurately; and (2) an arbitrary decision by the observer as to whether or not the recording procedure will be restricted to events only, states only, or a specific mixture of the two with the data being separated for the analysis. In many cases, it is possible to elect to record all behaviors as events without doing injustice to the validity of the observations, however, the reverse is not true for practical reasons, and considerable distortion will result from such an attempt.

State data, since it involves durations, can be utilized to examine such things as the "time budget" of a species or population and the proportion of time devoted to certain kinds of activity. Event data, on the other hand, really provides information about rates of occurrence of those behavior categories, or frequencies for a range of event occurrences. The former yields percentages of time, and the latter yields either percentages of all events or frequency rates per unit of time. It is important to recognize the distinctions between these two categories, as they will play important roles in the exercises appearing later in the book.

Altmann (1974) in her discussion of sampling methodologies identifies a number of recording procedures and evaluates them as to their usefulness in producing data that is valid for statistical analyses. Many methods employed among researchers prior to 1974 were considered to be, in essence, useless. However, a few processes were evaluated as being of limited use in particular research designs, and two—focal animal sampling, which is also known as continuous sampling, and scan sampling, variously known as interval sampling or instantaneous sampling, were given approval for general use while other methods were seen as constrained or limited to particular situations (see Table 2.1 for Dr. Altmann's recommendations). In the modern primatological literature focal animal sampling (fixed period of continuous observation and recording of a single individual or small group of individuals) and scan sampling (record of behavior states taken at some fixed interval for all members of a group) procedures are the predominant techniques. A subsidiary technique, focal time sampling, which might also be called focal animal interval sampling, was developed by Baulu and Redmond (1978) and has proven to be a valid mechanism. These sampling techniques are the bases upon which the observational exercises are constructed, and each is defined and discussed further in the introductions to each method.

Altmann provides a guide to the selection of sampling methods for various kinds of projects. Thus if the project is behavior oriented, *ad libitum*, all occurrence, and sequence sampling methods are better. If the project is individual oriented (that is, it requires identification of individuals and keeping separate behavior records for each), the focal animal and interval (instantaneous) scan methods are best. Only when there is a combined focus (e.g. we want to know about the grooming behavior of adult females) should the sociometric matrix method, or less desirably the one/zero sampling protocol, be employed (see Figure 2.3).

For each method, the capability of "event or state" recording and the most appropriate study problems are recommended. Hence sociometric matrix completion records only events and is most appropriate for a study of "asymmetry" or unequal frequencies between two individuals, and sequence sampling can record both events and states, but is only relevant to studies of why or how two or more behaviors occur in a temporal chain (what must come before or after a particular action). Other recommendations will be dealt with in specific exercises. The full set of Altmann's sampling recommendations is reproduced in Table 2.1.

Table 2.1 J. Altmann's recommendations for each sampling procedure (1974)

Method	State or Event	Recommended for:
Ad libitum	Either	Heuristic value only; suggestive; rare events
Sociometric matrix completion	Event	Asymmetry within dyads
Focal animal	Either	Sequential constraints; % of time; rates; durations; nearest neighbor relationships
All occurrences	Usually event	Synchrony; rates
Sequence	Either	Sequential constraints
One/zero	Usually state	None
Instantaneous and scan	State	% of time; synchrony; subgroups

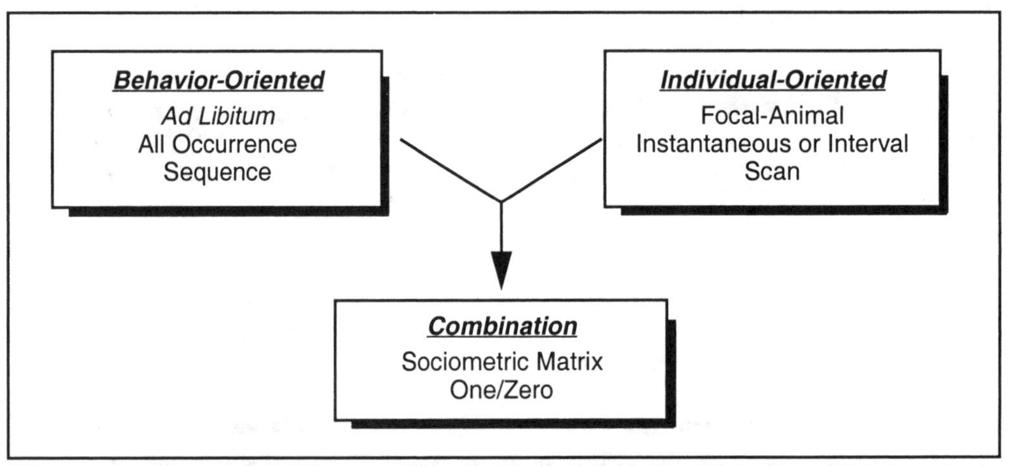

Figure 2.3 Altmann's 1974 suggestion for Sampling Method Selection. Which methods are the most appropriate depends on whether the research is oriented toward or focused on the behavior itself or the individuals or a combination of the two, such as "adult male aggressive behavior."

Figure 2.4 shows a hierarchy of sampling and recording rules in order to clarify the process of developing a research design. The first step is selection of a sampling rule, and while ad libitum and behavior sampling are available, focal or scan procedures are preferable. The second step is the recording rule, how the data is to actually be taken, with the choices being as a continuous record or through recording at intervals. If this latter choice is implemented, there are two methods available—instantaneous sampling and one/zero sampling (also known as Hansen frequencies). These are similar but display distinctive characteristics, as Figure 2.5 demonstrates. The difference is that for instantaneous sampling, *sample point data* is recorded *instantaneously at the end* of an interval, while for *one/zero sampling*, if a

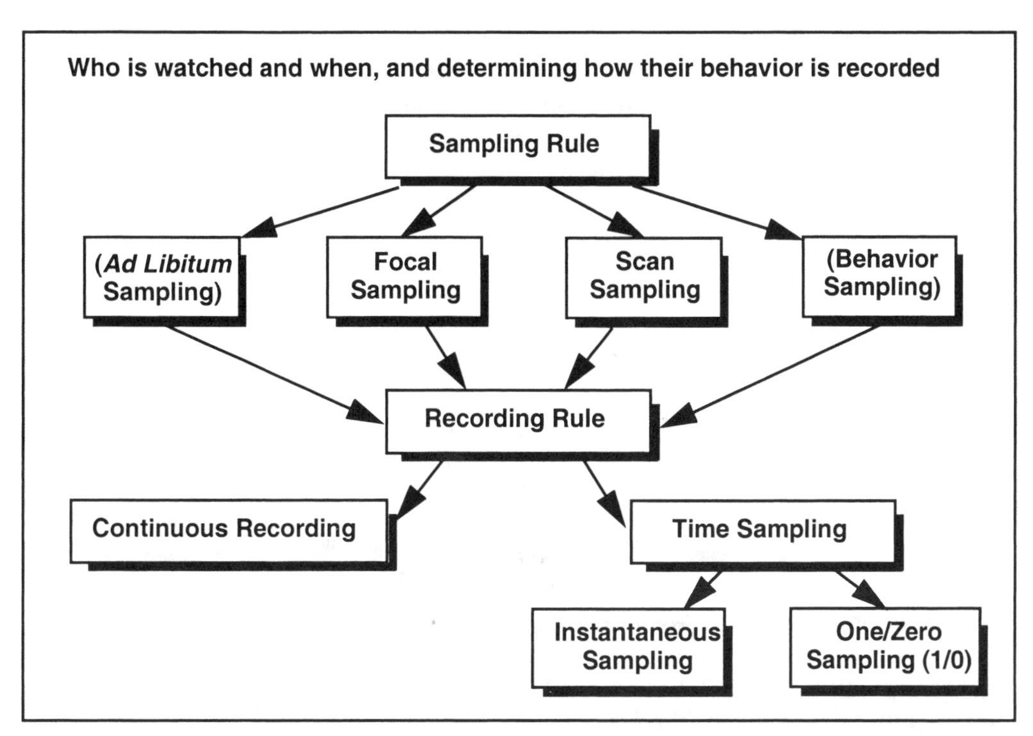

Figure 2.4 The hierarchy of sampling and recording rules (from Martin and Bateson, 1993:88). Martin and Bateson employ slightly different terminology from Altmann's. "Behavior Sampling" lumps together Altmann's (1974) sociometric matrix completion, all occurrences, and sequence sampling.

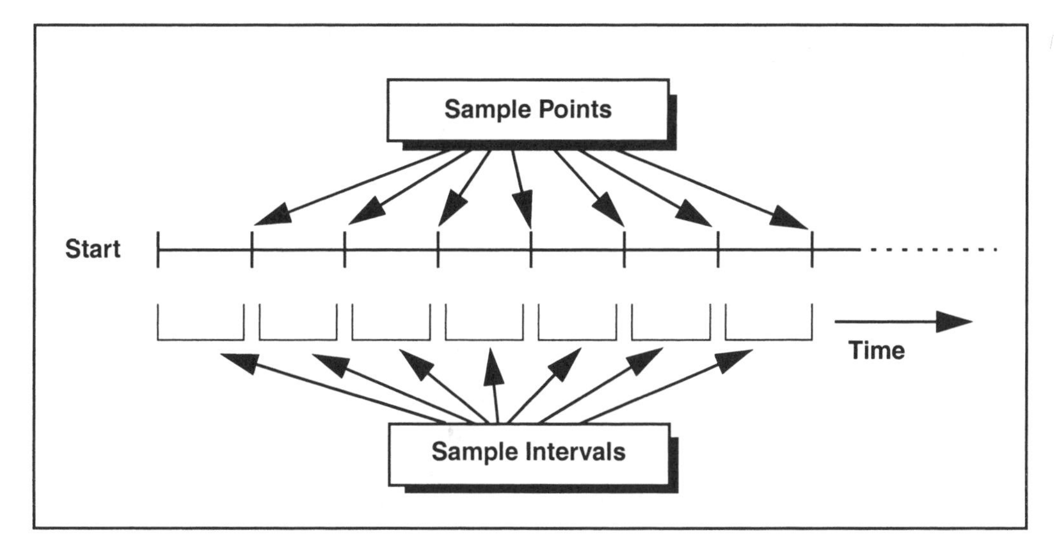

Figure 2.5 Relationship of sample points and sample intervals (from Martin and Bateson, 1993:89)

behavior occurs *during* the interval, it is recorded as a "one," and if it does not occur, a record of "zero" is entered. The rule for sample point records applies to all interval sampled methods except the one/zero procedure.

The sampling methods employed in any research project have a major impact on the manner in which the work will be done, and of equal importance, on the validity and usefulness of the data collected. It is of critical importance that care and attention to the details of sampling be incorporated into the entire scientific process.

RESEARCH DESIGN

The essence of research design, potentially an extremely complicated topic on its own (see Brim and Spain's *Research Design in Anthropology* [1974], Martin and Bateson's *Measuring Behaviour: An introductory guide* [1986, 1993], or chapter six of Lehner's *Handbook of Ethology*, 2d ed. [1996]), consists of: *defining a problem, forming a hypothesis about that problem, devising a method of study to provide data that has the potential to reject the hypothesis, and finally the execution of the study.* An important technical note must be placed here, and that is the explicit statement that "hypotheses" are always framed in pairs, one positive and one negative, *and all testing of a hypothesis is orientated toward the objective of rejecting a hypothesis, normally one of the pair.*

This means that in evaluating a pair of hypotheses, it should become possible to reject one of the two and to fail to reject the other. *This does NOT mean that the alternative hypothesis is proven to be true*, only that it has failed to be rejected and can be provisionally accepted, pending further testing. To avoid problems, the hypotheses being tested should be logically constructed so that there is no logical or theoretical confusion surrounding them. The ideal formation should be a logical binary "yes or no," "it is this way or it is not this way," and if the hypothesis can be broken down to component hypotheses, then these are the ones that should be subject to testing. In order to conduct this testing, it is necessary to establish some relevant variables, evaluate what the variables measure, and how precise they are at doing so.

Variables and Measurement

Variables and measurement constitute both simplicity and inordinate complexity in the study of behavior. A "variable" is simply something that varies—in frequency, in duration, in time of occurrence, in location, with social situations, in form of expression, and so forth. Measurement mechanisms associated with these therefore come in a variety of forms correlated with these variables. But it is quite easy to set up a variable and measurement structure that will produce meaningless data. As the early computer scientists had it, there is a "GIGO" rule attached, "Garbage In, Garbage Out," and no amount of statistical juggling can turn meaningless data into useful models of behavior.

Variables fall into four distinct categories of measurement schemes, or more familiarly "scales." The scales of measurement are *nominal, ordinal, interval,* and *ratio.* It is normally of importance to be aware of the scale of measurement related to a variable as this often restricts the choices of appropriate statistical analytical techniques.

Nominal. The measurement units within the variable (for example *behavior*) are a set of named qualitative units sometimes conceived of as *attributes* (e.g. locomoting, feeding, grooming, playing, etc.) and that can only be counted. The basic data derived from nominal level data are frequencies and relative frequencies (see chapter 4), and the kind of statistical procedures appropriate to the data is limited. The so-called *nonparametric* statistical tests are most frequently employed.

Ordinal. This scale is similar to nominal except that the categories are ranked into an ordered sequence. This ranking may relate to either a quantitative or qualitative sequencing. The distinction between the categories is not determined by a fixed distance from one to the next, but derives only from the order, in that one comes before another. In other words, the scale is one of relative placement in a sequence.

Interval. This scale differs from ordinal in that the categories are both ranked and separated by a fixed interval along a continuous measurement scale, but there is either no zero point, or it is arbitrarily assigned. As has often been pointed out, the most commonly employed temperature scales, Fahrenheit and Celsius, are interval in nature since their "zero" points are essentially arbitrary.

Ratio. This is similar to interval except that the zero point is known. The most familiar ratio scales are time and distance, since both have a recognizable start point and display continuous interval measurement that is ordered in sequence.

Studies of behavior regularly employ variables of the nominal, ordinal, and ratio scales, but rarely is an interval scale employed. The only way to determine what scale is employed in a study is to examine the raw data and note how it was recorded. A key or codebook often will provide this information directly, but it can be determined by inspection. If the record consists only of named units, it is nominal; if there is some added information that relates to order, it is ordinal; if the time that the behavior is recorded is noted, it can be considered as interval; and finally, if both time and duration are recorded, the scale is ratio.

All scales are subject to issues of accuracy and precision. *Accuracy* is a statement about how closely the variable measures the issue related to the hypothesis. To employ a sporting analogy, it is an Olympic marksman being able to hit the center of the target. *Precision* is a statement about how consistent the observer is with the recording of the behavior categories. In the sporting analogy, it is the marksman being able to *consistently* hit the center of the target. An illustration (Figure 2.6) derived from the first edition of Lehner (1979) shows this relationship. The concepts represented here are that a bias may shift the behavior record off the target (poor accuracy). Usually biases can be accounted for and corrected during an analysis, much as the marksman adjusts the gun's sighting mechanism. The failure of precision (i.e., when the form of the behavior record is variable or inconsistent) can lead to unrecoverable errors and distortion of the data set. Observers need to strive to be *both* accurate and precise.

Example. As an example of research design, consider the phenomenon of male dominance hierarchies—the "pecking order" seen among terrestrial old world monkeys such as the baboons (*Papio cynocephalus* also known as *Papio hamadryas* in recent taxonomic literature). It is obvious, after a short period of observation, that a hierarchy ranking males against each other exists, but what factors can be used to measure male dominance? Many factors are candidates—fighting skill, body size, ability to make favorable alliances, the favoritism of the group's females, intelligence, or lineage relationships (that is "who" one is related to within the group, and what status he occupies). What behaviors give clues about these factors? To select just one—favoritism of the females for example—can we see in the behavioral pattern any data leading to the conclusion that "a male has to be a favorite with the ladies in order to achieve high rank"? The alternative hypothesis, called the null hypothesis, would be "a male does *not* have to be a favored partner of the females in order to achieve high rank." One factor that might provide data to evaluate these hypotheses is the amount of grooming a male receives and from whom it is received. It thus becomes possible to construct a study in which one focuses upon the males in a group and records the amount of grooming each receives and from which females, and which females the male grooms in turn. One can then test the hypothesis that the male who receives the most grooming, measured by the number of females participating in grooming him, or by the total time of grooming, should be the male who is the most dominant in the male hierarchy. Once the study is done and the statistics calculated, the male with the highest amount of grooming or the largest number of females grooming him, should be the most dominant in the hierarchy. If this is the case, the hypothesis fails to be rejected, and hence can be provisionally accepted, but if a different male is most dominant, then the hypothesis is rejected and the null is provisionally accepted.

Obviously, the data leading to affirmation or rejection of that hypothesis are representative of only a single variable, and a better design is to utilize a number of different measures related to different factors in order to reach a valid conclusion. Once the problem is set, the hypotheses formed, and the research design established, the actual conduct of the study and the analysis of the data may turn out to be anticlimactic. In fact many studies come close to being confirmations of "intuitive" perceptions formed by the observer before any planning is undertaken—in one sense this is a perfect model of the scientific method.

A preliminary consideration that saves a great deal of pain and irritation later on is some attention to the subsequent data analysis. That is, in the

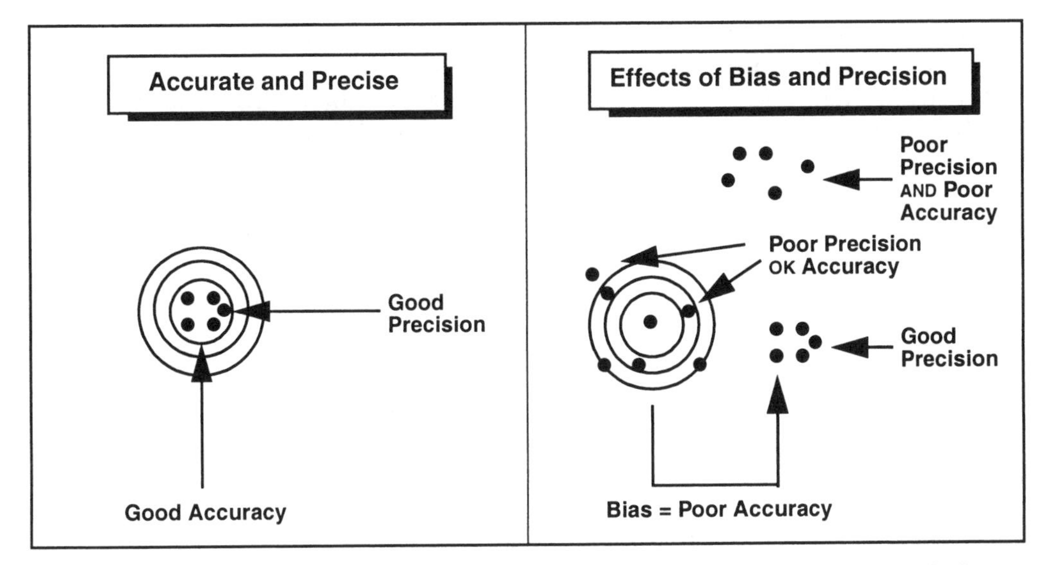

Figure 2.6 The relationship between accuracy and precision (modified after Lehner, 1979:128). The left image shows a good centering on the target (accurate) and a small grouping (high precision). The right image shows the possible combinations of errors. A bias (error of slight adjustment) can lead to a case of poor accuracy but retain good precision. The sights may be set correctly but the operator cannot achieve consistency, leading to adequate accuracy but poor level of precision. Finally both types of errors may be present, as in the upper right grouping which is both off the target (poor accuracy) and widely dispersed (poor precision).

preparation of a research design, it is appropriate and necessary to consider the types of data summarizations and statistical tests that will be used later on, and adjust the design and data collection protocols to conform to the analysis requirements. This may merely involve including appropriate spaces on check sheets for summing data. As noted earlier, it is good practice to perform some preliminary analyses after an initial round of data collection; check to see if the data are applicable to the problem, and revise the research plan, check sheet, and/or hypotheses as necessary.

Drawing Conclusions

Students are often intensely concerned with finding some form of "absolute truth" in scientific endeavors, and sadly they must be advised that this objective is impossible. All that science can provide is progressively more accurate representations of what is out there in the real world. In any scientific report stemming from any scientific observation, it is important to recognize what conclusions are possible from the data collected, and what are not. To be exact and precise, one can restrict the report to a straightforward listing of the hypotheses accepted and those rejected, but it is the norm that a degree of interpretation of the results is allowed. However, it is necessary to consider carefully how far this interpretation may go. While it may be possible to conclude that "a male has to be a favorite with the ladies in order to achieve high rank," the causality of the relationship is not defined in the test of the hypothesis. That is, what has been found is a correlation between male rank and male acceptance by the females; there is no causality involved. Explicitly, we *cannot* say that males have high rank *because* they are favored by the females, nor can we say that males are favored by the females *because* they have a high rank. Other hypotheses and testing are needed to provide such an insight into the relationship. Conclusions must be drawn from the data collected and analyzed. If conclusions cannot be drawn, either the researcher fails to recognize what has been found, or the research goes beyond the realm of science into the world of speculation. Care and caution are clearly indicated in the activity of drawing conclusions.

OBSERVING SCHEDULES

Many of the exercises in this volume require the use of some form of an observing schedule. The plan for the actual conduct of research work is normally prepared in advance and must enable the worker to complete the study efficiently and in conformity with the requirements of the research design and of the analysis procedures to be used. An observing schedule is somewhat like a bus schedule; it is laid out in advance and designed to get to all of the stops on the route at fixed times. However, like bus schedules, they tend to get out of control, either running late or missing planned stops.

There are two main kinds of observing schedules, the *fixed format* and the random sampling schedule, as well as two common variants—*haphazard* and *opportunistic* (see Lehner 1996:144–147 for discussions of these), and they both have their useful aspects. They are often linked to the size of the social group under study and to the time span over which the study is conducted. The fixed format schedule is useful for small groups, short duration studies, or for specific experimental sampling of a particular set of members from a larger group. Random sampling is always to be preferred where the data will be analyzed using inferential statistics.

Fixed Format Schedules

Fixed format scheduling is a very deliberate process. The observer sets up a working program to collect data from individual subjects at specific times for specific durations. The planning of such a schedule must take into allowance such factors as the climatic conditions, the availability of the subjects, and the requirements of the research design. As an example, during the summer of 1982, Paterson studied the postural and orientation behavior of a set of ten adult male Japanese Macaques (*Macaca fuscata*) at the South Texas Primate Observatory near Dilley, Texas. The research design required a full-day sample for each male, the research period was limited to thirty-two days, and the daytime maximum temperatures during the period were expected to peak in the 40–45°C range. In consequence, the use of a fixed format schedule allowing for the collection of four or five hours of data each day was utilized, and a schedule plotted out for the period specifying which subject was to be the focus of attention for each work hour (i.e., "if this is Tuesday July 17th at 1100 hours the focal subject is male #134"). *Such rigorous schedules must make allowances for rest periods (for the observer), seek times (the time it takes to find the subject assigned to that sample period), and the possibilities of the climatic conditions making observation impossible.* Fixed format scheduling is also the easiest to use for the conduct of studies at a zoological facility.

Random Sampling Schedules

Random Sampling schedules are necessary under two circumstances:

1. when the study group is large (and all individual members are being studied over an extended time period), and
2. when the data set is to be analyzed by inferential statistics.

In order to conduct a random sampling, the observer must utilize a procedure that will acceptably produce a random allocation of times to observe particular animals, or of particular animals to observe at specific times. Computer programmers can produce code that will do this job automatically for them, but one of the most commonly utilized methods is also the simplest—drawing names or times out of a hat. This process can be conducted in two distinctly different but equally effective ways:

1. The observer creates a set of slips of paper or chips with individual animal identifications on them, and then at the beginning of each new sample, merely draws one slip. This slip identifies the next subject to be studied AND IS THEN RETURNED TO THE DRAWING BOX OR HAT. This last act is a vital requirement in order to guarantee that each subject has exactly the same chance of being drawn for the next sample period.
2. The observer sits down the evening before the work day and proceeds to do all of the drawings for sample periods at one time, but must be careful to return drawn slips to the box after noting the subjects' identity. The result is a listing of the order of observation for the subjects of the next day. One extra requirement of this process is that a list longer than the intended set of observations be constructed so that if a particular subject cannot be located within a reasonable time, the observer can go onto the next sample, adding the first of the reserve list to the end of the day's sample series.

An alternative method is to create a listing of the subjects to be studied, and to then perform the drawing process with slips containing times for the observation sample. This alternative is generally less desirable than option two since it results in a decreasing pool from which to draw, thus violating a true random distribution.

A very common alternative procedure for either of the above processes is to substitute a set of dice for drawing slips. This does however require that each subject be allocated a number, and that each number is capable of being generated by a random throw of the dice. Those who are familiar with fantasy role-playing games might consider the use of more complex dice than the traditional cubic form and may be able to find dice with appropriate numbers of sides for most occasions and group sizes. However, a more appropriate procedure is to employ a stratified partition method such as that detailed in exercise 4.

The apparent inequities of the random sampling process, and the inevitable cases where the same subject is selected for several consecutive sampling sessions, may lead one to conclude that the selection process is biased; however, all of the factors balance out if the study is of sufficient length. Thus, a secondary requirement is that the study must be long enough in duration that the random selection process can have the opportunity to bring the sample sizes for all of the subjects up to approximately the same level, over the same study hours. This allows the research process to reach a truly random sampling of the behaviors of the group. However, for most short-term projects, and the exercises in this book, a fixed observing schedule can be more effective.

Setting Up Check Sheets

While there are a number of different ways to record data, the oldest and most commonly utilized is some form of the check sheet. A check sheet may range from a simple blank page to a highly structured instrument geared toward precise and complex observation. As Hinde (1973) has said, "While the design of a check sheet seems a simple matter, experience shows that there are snags which may handicap research efforts." Every researcher and every project are different, and thus "every check sheet must be designed with an eye both to the problem in hand and to the idiosyncrasies of the observer" (p. 393). As a result of these features, there can be no "standard" check sheet; however, there are a number of principles that can be employed to guide the design of a functional sheet related to a research question.

A check sheet always requires two main components:

1. *Data management and control areas.* These are normally placed at the top of the sheet and provide the information about the observer, the date and times of observation, special control factors, and other items that may be relevant to the analysis.
2. *Data collection fields.* These cover the remainder of the check sheet and should be constructed in a form related to both the sampling protocol and the recording method.

A simple check sheet may be constructed with only a limited amount of control data, and a minimum of categories, as in the example in Figure 2.7 from Bakeman and Gottman (1997). The control data consists only of the observer identity and the date and times of start and stop. Because the sheet was designed to deal with a simple categorization of positive and negative behaviors in human children at a play school, there are only three columns and the record consists solely of a tally count of these as perceived by the observer. Clearly this simple sheet provides a very limited amount of both control information and data. It has a very limited utility and can provide only a frequency count of "hits," "quarrels," and "aid requests" by an unknown number of individuals during the observational time. The most useful data coming from such a record would be a rate of occurrence.

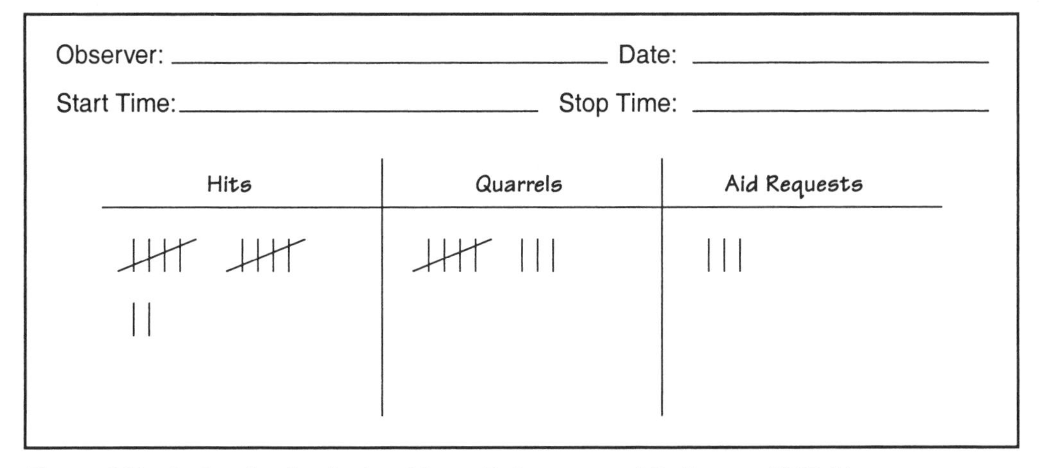

Figure 2.7 A simple check sheet from Bakeman and Gottman, 1997:41.

Hinde (1973) presented an example of a significantly more complex check sheet in which there is a column for time, another for the identity of the subject chimpanzee, and five sets of columns for tallies or notes on behaviors that might occur, plus a larger column for comments. This is typical for an actual operational check sheet (see Figure 2.8).

Time control is often a requirement of modern research designs, either as regular intervals, or as a continuous record. While examples of appropriate check sheets for these will be found in the appropriate exercises, it may be worthwhile remembering an older approach sometimes called lined paper recording. This is actually a manually operated version of the data recording style generated by event recorders. The structure of such a check sheet is straightforward. The first column is for time, at appropriate intervals, and the remaining columns are coded for particular behaviors. Thus, the sheet lists each behavior horizontally above a column of boxes that coincide with time in intervals arrayed vertically. The observer may use check marks or simply draw a line in the appropriate column from the start of the behavior to its end. Such a check sheet would look like Figure 2.9.

Interpretation of the data in such a sheet is simple. Behavior "A" occurred from time 1 to 4, and time 11 to 13, in both cases associated with "Sit" posture,

TIME	CHIMP	arrival	CALL	FACE	charge	hair out	bipedal	slap	stamp	drag	branch wave	throw (object)	drum	CHASE	CONTACT I	CONTACT II	ARM THREAT	APPROACH	GROOM	BY	HUNCH	EMBRACE	KISS	EXTEND HAND	TOUCH	SEX	BOB	CROUCH	PRESENT	CLOSE FACE	AVOID	NO RESPONSE	BOTTOM CONDITION	COMMENTS
1200	HM	✓	Ph		✓	✓		1				2	3																					Throws rock
1230	FG	✓	Ph		✓	✓		2		3	1																							Drags branch
	SW		Sc																												FG			up tree
1238	FG					✓												SW	SW	SW														
	SW		Pg																FG	FG									FG					
1300	MK				✓	✓								SW							SW					SW								See sex
	SW		Pb															M^3K								M^1K			M^3K		M^1K			up tree then approach

Figure 2.8 Check sheet for recording particular types of behavior when they occurred in chimpanzees arriving at the Gombe Stream Research Center (Hinde, 1973:402 based on a personal communication from Halperin).

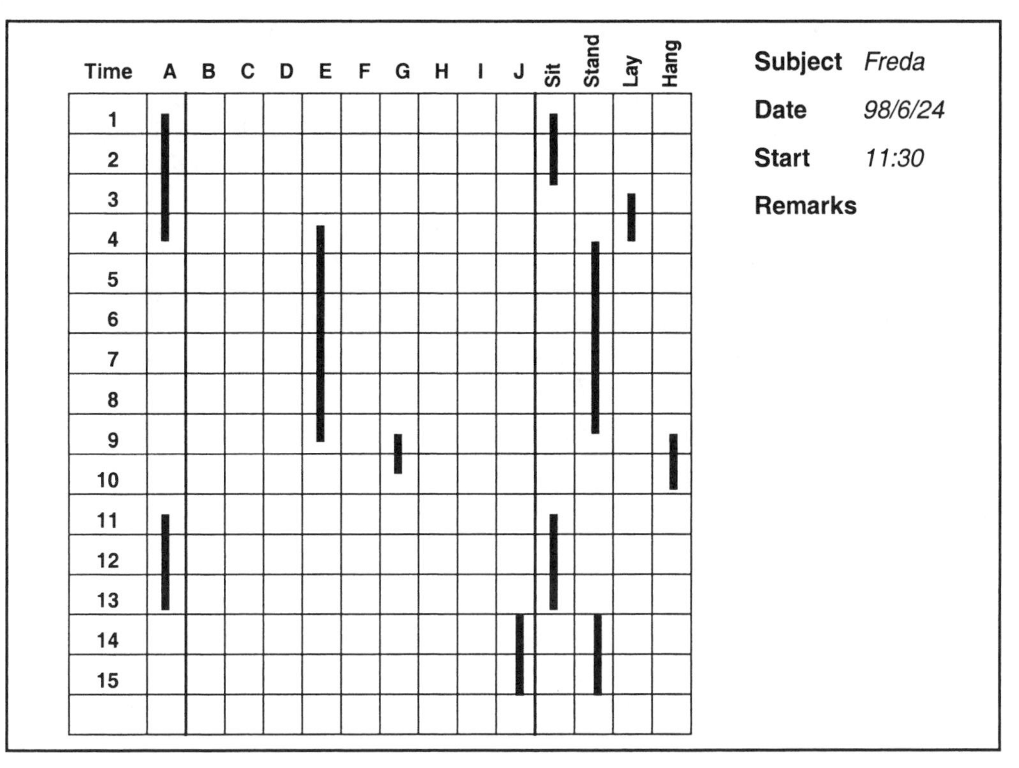

Figure 2.9 Lined paper record check sheet for ten behaviors and four postures.

but also with "Lay" from time 3 to 4. This layout could be employed for continuous time recording, through leaving the time column blank and writing times of behavior or posture changes, or for interval time recording with regular times printed in. It is, however, limited to a preconceptualized and predefined set of behaviors and/or variables. Data falling outside of these parameters can be added only under the informal "remarks" area. Every observer will eventually need to design and test check sheets for particular observational problems, and one of the exercises will engage the student in this activity. However, for many of the other exercises, sample check sheets are provided in order to spend time observing rather than constructing check sheets.

The "Out of Sight" Problem

In any observational study conducted at a cage enclosure in a zoological park, one of the consistent problems that surfaces is what to do when a subject of a focal animal (continuous) sample moves into an unobservable part of the enclosure. There are variable policies in zoological parks, sometimes animals are locked out of portions of their habitat for substantial periods of time, while in other cases, the subjects may have complete access to all areas—on and off display—throughout the day. It is generally appropriate to inquire as to the policies of the keepers and to modify research designs to accommodate these variations.

Where it is possible for a subject to escape from the observer's view, it is necessary for the observer to include a category of behavior labeled simply "out of sight." This naturally leads to another situation requiring a policy decision within the research design: At what point does one terminate and discard a sample due to an extended "out of sight" condition. This is an arbitrary decision on the part of the observer, but it makes little sense to include samples in a study where the animal is out of sight for more than half of the sample.

There are six mechanisms to deal with the problems arising from the out-of-sight condition. Four of these are discussed by Lehner (1996), but as he clearly notes "*none of these methods is truly valid*," and I have watched statisticians cringe at the suggestion of using one or more of the techniques. Note that all introduce some form of bias into the data. Where the out-of-sight period is long in relation to the normal durations of the behaviors, Lehner suggests two variants:

1. Deletion of the out-of-sight time from the overall sample. (The length of the sample in this case is reduced, and many statistical tests are not able to handle data in noncomparable blocks.)
2. Timeout. Suspend observation for the period that the animal is out of sight, resuming the sample on its return. (Again this is a procedure that can bias the results of statistical tests in unknown and unforeseeable ways.)

Where the out-of-sight periods are of short duration, Lehner provides two alternates:

1. Allocate the same behavior seen at the time of going out of sight to the whole period (i.e., if the animal is "foraging," then the category of "foraging" continues until it returns to view).
2. Allocate the same behavior seen at the time of returning to view to the whole period (i.e., if the animal is "locomoting" upon return to sight, then the out-of-sight period is assigned "locomoting" as well).

Lehner suggests that if the behavior is the same for both methods, the allocation of the out-of-sight behavior is more valid; however, this may not necessarily follow. Lehner provides only a relative guideline, and no fixed rule is readily available.

The two other solutions are more direct but also have statistical or research consequences. One solution is to utilize the category of "out of sight" as a regular behavioral unit and employ it within the statistical analysis. This may have the simple advantage of not biasing the statistical procedures, but it clearly intrudes a substantial element of uncertainty into the data. Treating "out of sight" in this fashion can have untoward research effects that may be variable across the diurnal cycle for instance. This yields a distortion of the data in spite of being in statistical compliance.

The final option is to establish a protocol rule dealing with how much out-of-sight time can be tolerated. It may be suggested that an absolute limit is one third of the sample time as the point at which the sample should be discarded. That is, if the out-of-sight time exceeds five minutes in a fifteen minute sample, it is discarded and another sample started. However, it might be decided that out-of-sight conditions are not representative of the behavior of the subject animal, and the tolerance for this category may be set at zero, hence discarding a sample as soon as the subject leaves the area of view. This results in a larger number of discarded samples and the collection of a limited number of valid samplings during each working session. These are decisions that each observer has to make, but the pattern must be consistent throughout the data-collection phase.

Observers may feel guilty about throwing away samples in this fashion, but it is important to realize that it could be far worse to include all the out-of-sight records and thus wind up biasing the analysis.

Each observer will see the occurrence and effects of out-of-sight records in different ways, and consequently, each researcher's protocols and research design will be different. But it is important to include a protocol, normally some combination of the ways to deal with the out-of-sight problem in the research design and know what to do when the situation inevitably arises.

Selection of a Subject Species

One last issue in research design must be dealt with before proceeding, and that is the issue of "choosing" a subject species. This may be perceived as irrelevant by those individuals who are locked in on a specific species as their focus of interest. They are restricted in their selection of research questions to those that actually can be conducted with their favored species. For those who have not homed in on a subject species, there are four basic considerations in the selection of a species for a particular research question, and these have been succinctly gathered together by Lehner (1996).

Suitability. The August Krogh Principle asks, "Is the species suitable for the concept being studied?" In other words, one first defines the research problem, and then searches for a species that displays the relevant kind of behavior and, furthermore, does it frequently enough to provide adequate data. It must also be possible to observe the species, and the behavior of interest for it will do no good to the project if the activity is conducted out of sight.

Availability. Can you easily go to the subjects or have the subjects brought to you? Two primary conditions exist: If the species is not locally available and must be studied in the field, what are the political issues? Is it possible to conduct research in the country or countries where the animals exist? What restrictions or permits have to be obtained, how are they administered, are fees payable for the permits, can the permits be obtained before arrival in the country? Any and all potential problems with the legal and political establishments must be considered. Then there are direct observational issues: Is the species accessible; can it be easily observed; is the species nocturnal, diurnal, or crepuscular; do your personal habits fit with the animals' schedules? If your project requires that the subjects be maintained in a laboratory, then a number of quite different problems arise. Is the species on one of the endangered species listings? If it is, there is unlikely to be any way that you will be permitted to obtain subjects without long and complicated legal procedures. If not, can it be easily captured, can it be easily maintained in the laboratory, does it require unusual dietary materials, and so forth?

Adaptability. Is the animal able to adapt to captivity? Are there any special caretaking needs that may interfere with the experimental protocols? Are your lifestyle patterns compatible with your subjects'?

Available data. What is already known about the species? Perhaps your research question has already been tested. Is there a good background data set? If there is, will these data help to answer the foregoing questions and anticipate problems?

Decision time. The positives and negatives related to this set of four characteristics must be added up and an evaluation of the financial commitment involved factored in before choosing to accept or reject the species as the subject for your investigation.

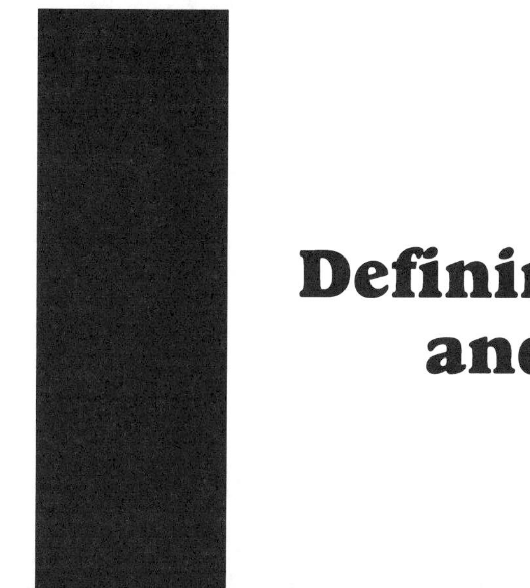

3

Defining Behavior and Coding

Definitions of behavior abound in the literature. They run from a simplistic "Behavior is what an animal does"; to Hinde's (1973) statement, "Behavior consists of a complex nexus of events in time"; and even to Heraclitus of Ephesus's famous philosophical dictum "you can never step into the same river twice." What all such statements are attempting to do is provide some empirical, rational mechanism for describing and categorizing acts, actions, and interactions of our subjects.

As Hinde (1973:393) goes on to say: "Before it can be studied it must be described, broken down into units suitable for study, and the units classified into groups according to common properties." This statement clearly implies that there is a difference between "behavior" and "behavior records." The former is truly the stream of operation of the organism; the latter is the record produced by the observer. To some extent this may seem like a restatement of the philosophical conundrum "If a tree falls in the forest and there is no one to hear, does it make a sound?" but it is not. What it really implies is that the organism is *independent of the observer*, it performs (sometimes it is said to "emit") a stream of behavior, *not all of which is visible, or audible*, and the *record* produced by the observer has been filtered through the neural system of a different organism. The record is therefore an abstract of what the observer saw, which in turn is an abstract of what the organism did.

Often it is assumed that the definition of behavior is an activity performed at or near the very beginning of a study. Unfortunately, defining behavior requires a degree of experience with the subject species and their activities before the observer is competent enough to undertake the description and categorization of the sequences into a definition. In addition, not all behaviors are observed during the initial phases of a study, and new behavior definitions will need to be added, but at progressively less frequent intervals until the project ends. In ordinary terms, it may be possible to record a very large portion of an organism's behavior repertoire in ten to a hundred hours, it may be possible to record 99 percent of that repertoire in ten thousand hours, but it is unlikely that the book can ever be closed on the behavioral catalog.

Behavior and Perception

It is important for the beginning researcher to recognize and remember that *a behavior record of the activities of subject animals is a mental construct of the observer, and as such is subject to bias and distortion that may divorce the reality of the subject's action from the perceived and recorded observations*. For an observer to record behavior that is "seen," a moderately complex neural pathway is traversed by the information. The image is then compared and contrasted, and recognition, pattern-seeking functions employed to "interpret" what the image is and associate it with a meaning. *The most dramatic insertion of biases and selective perceptions take place at this level.* In a sense, people may see only what they want to see, even when they are trying to be "objective." It is also important to recognize that this situation can become significantly worse if the observer is fatigued.

In order to correct, partially, this set of problems, a threefold level of description for behavior is presented as one mechanism to aid in the control of perceptual biases.

Levels of Behavior Description

Some researchers describe behavior at "atomic," "molecular," and "integrated interaction" levels. Other researchers have made their level separations somewhat differently, utilizing only two categories—molecular and molar (Coelho and Bramblett, 1990; Sackett, Ruppenthal, and Gluck, 1978). For purposes of comparison, "molecular" in the latter use approximates what is referred to in the former as at the atomic and a substantial part of the molecular levels, while the "molar" includes most of the molecular level and some higher aggregations. The separations as they are presented here reflect a distinction between purely anatomical movement descriptions (atomic) and the units into which these can be aggregated to form recognizable behavioral units (molecular), while the third level is partially an aggregation of molecular units and partially what Hinde has labeled as a description by consequence.

Atomic description can be an essentially unbiased level that endeavors to record the fine details of behavior through accounts of the individual muscle or segmentary movements and skeletal actions. This level of description records purely the actions of components of the subject animal; it is not concerned with the activity of the whole subject, but only its parts. Atomic-level description is extremely difficult work that also produces an enormous mass of data. Much of this mass is of little direct utility in understanding the larger-scale activity and interactions of a species; however, it is the case that atomic-level units of action are straightforwardly concentrated into consistent larger patterns that can be labeled as molecular units of behavior.

Molecular behavior units are generally easily recognizable categories, and similarly appear to have identifiable functions such as serving as a signal.

These molecular units of behavior are subject to some encroachment of biases but on the whole are much more useful to the observer than atomic records. While atomic units of behavior might describe a human smile as "*progressive stretching of the orbis oralis muscle towards the auricles by the levator oralis muscles, accompanied by slight parting of contact between the surfaces of the lips*," we tend to perceive the action as simply a "*smile*," and, in an anthropomorphic interpretation (that is, one specific to our own species), to impute various "*friendly*" emotions to the action. In many other species this action pattern would be considered as threatening, appearing as an "intention to bite" or a variant of "tense mouth face" (described later as a low-level threat signal). Molecular behavioral units are such activities as walking, resting, feeding, drinking, and especially for primates, facial expressions (see van Hoof's 1967 catalog of primate facial configurations, Figure 3.1).

The third and highest level of categorization, the "integrated interactions" are often, but not necessarily, formed from sequences of molecular units. In many cases these behaviors are definable by the context in which they occur and are what Hinde (1973) called "described by consequence," that is, they are defined by their effects. For example, "approach" behavior has the effect of decreasing distance between two animals regardless of the mode of locomotion, and "supplanting" may be defined as the approach of one animal followed by the immediate departure of another, thus "supplanting" retains a suggestion that the approach of one

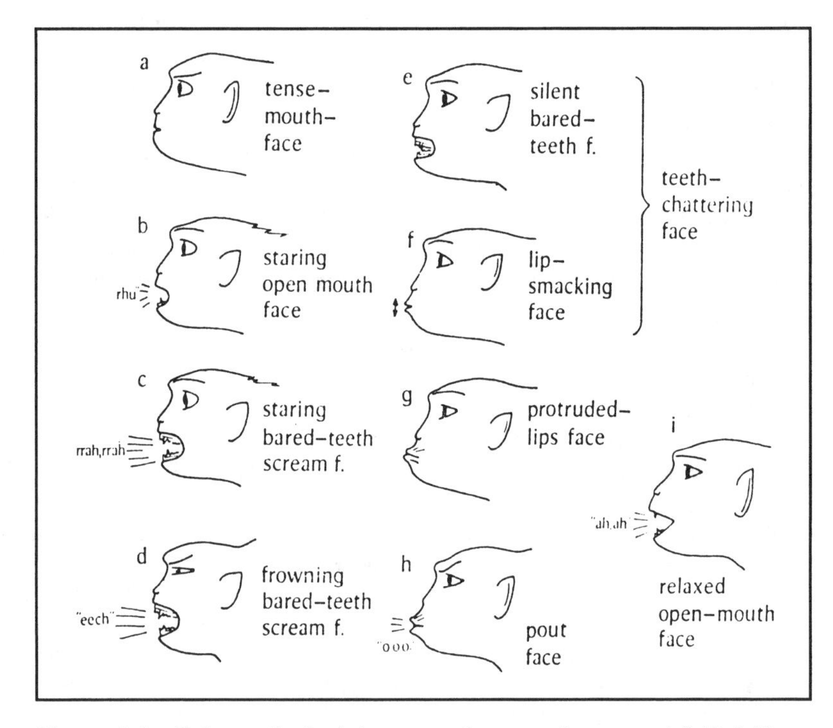

Figure 3.1 Schematic facial expressions and names J.A.R.A.M. van Hoof from *Primate Ethology*, Desmond Morris, ed. 1967: 64.

caused the departure of the other. Similarly, "avoidance" can be defined as the leaving of a location by one animal before the approach of another animal is completed. Again, "avoidance" retains a suggestion that the action is performed in order to "avoid" having contact with the approaching individual. The three terms are all related, and while there can be some difficulty in separating supplantation from avoidance, a "timing" judgment can be applied. If the approached individual is resistant to leaving and the approacher stands or sits next to the approachee, it will probably result in supplantation being recorded, but if the approachee leaves before the arrival of the approacher, then it is clearly avoidance.

Behavior Classification

Prior to embarking upon the development of formal definitions, the normal process requires some thought and attention to the classification of behavior, and the development of behavioral units. An initial turn around the zoological gardens and a few minutes of *ad hoc* observing often suggests that this process is relatively simple and easily accomplished. However, as Stuart Altmann (1968) has noted: "categorizing the units of social behavior involves two major problems: when to split and when to lump." Hopefully the boundaries between behaviors developed by the observer will reflect the splitting and lumping that the subjects themselves do rather than arbitrary acts of the observer. As is common in primatology and in ethology, there are an number of ways to classify behavior. Scott (1950) developed a nine-category list of adaptive behavior, as seen in Table 3.1, in which the categories are largely self-explanatory. A rather more complex and hierarchically structured categorization of behavior is the scheme developed by Delgado and Delgado in 1962. Their work was conducted with captive primates and linked to an explanatory model of behavior. Figure 3.2 is a block diagram of the hierarchical arrangement of their categories. The distinction between simple and complex is at the top of the hierarchy and reflects to some extent

Table 3.1 Scott's classification of behavior (1950)

Ingestive	eating and drinking
Investigative	exploring social, biological, and physical environment
Shelter-seeking	seeking out and coming to rest in the most favorable part of the environment
Eliminative	behavior associated with urination and defecation
Sexual	courtship and mating behavior
Epimeletic	giving care and attention
Et-epimeletic	soliciting care and attention
Allelomimetic	doing the same thing, with some degree of mutual stimulation
Agonistic	any behavior associated with conflict, including fighting, escaping, and freezing

a distinction between behaviors that could be lumped as atomic and molecular in the former with many integrated interactions placed into the latter. The breakdown in the simple units to individual and social is paralleled in the next lower level with both being separated into static and dynamic. These two categories separate what we may refer to as postures from active behaviors. The final hierarchy level under dynamic indicates whether the active behavior involves only a part of the organism (localized) or the movement of the whole entity within the environment (generalized). The complex behavioral units separate out behaviors that occur simultaneously either within one organism or coordinated between several individuals. Walking and talking, or the flight of a flock of birds, would be an example. It also separates sequences of behavior, a topic that will be examined further in the introduction to the applied project on sequential analysis. Delgado and Delgado also separate the category of syntactic behaviors, which are context dependent (i.e., the significance of the action depends on the circumstances in which it took place). Finally, the category of roles is used to point to the manner in which the action is undertaken. The example of grooming behavior in the following section is one behavior that can be categorized under roles and can be assessed as either active or passive.

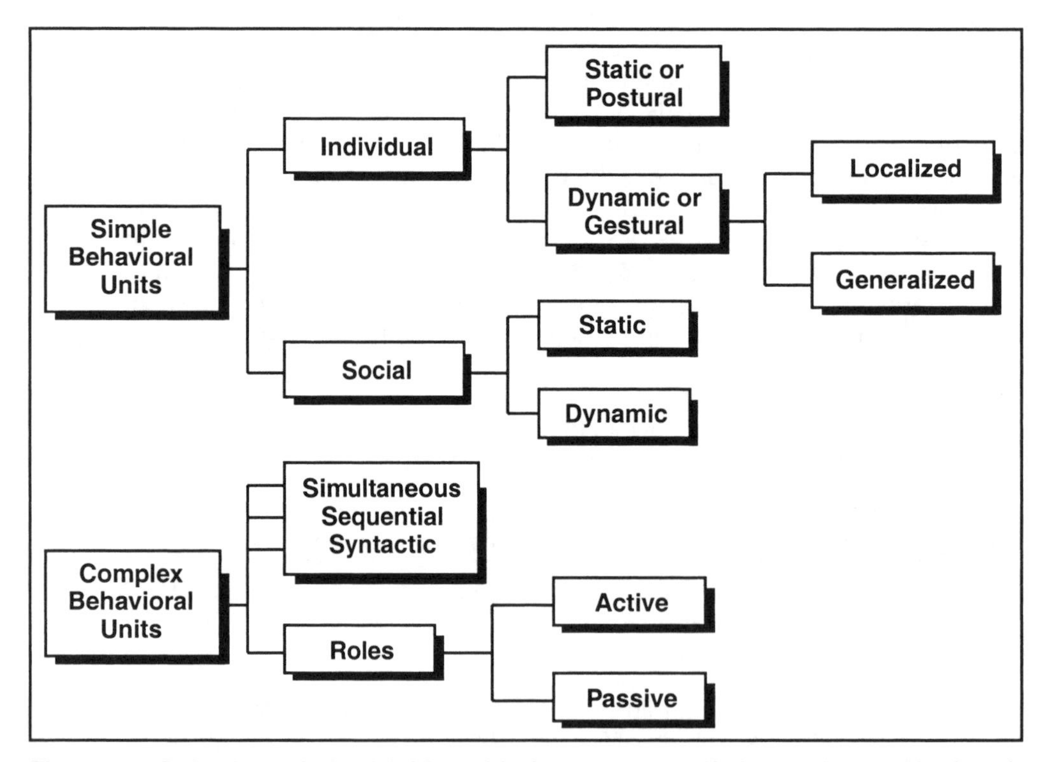

Figure 3.2 Delgado and Delgado's hierarchical arrangement of behavioral categories (1962).

Formal Definitions of Behavior

In an ethogram (see exercise 2), formal definitions of behaviors are required, but the amount of detail employed in a definition is variable. This can be clearly seen in the variations found within the Gorilla Ethogram collection compiled by the Gorilla Behavior Advisory Group of the Gorilla Species Survival Plan Group. This collection is available on the Web from PIN (Primate Information Network) at:

http:/www.primate.wisc.edu/pin/gorilla.txt

As a general format for a formal definition, it is suggested that a clear description employing as many atomic-level descriptors as is appropriate be generated. It is also a cardinal rule of philology that the definition may *not* include the term being defined. That means a definition of "sleeping" as "being asleep" or "the activity of sleeping" is not acceptable; however, one such as:

Sleeping remaining in a relaxed posture with closed eyes, regular breathing, and apparent unconsciousness

is a good functional definition. However, it must be acknowledged that there may be variations around this definition; for instance, some humans are able to sleep with their eyes either half or fully open (the key then becomes the absence of the blinking normally seen in conscious individuals).

A second requirement of behavior definitions is that they be both mutually exclusive and extensive. This means that overlap between behavior categories is undesirable, and the behavioral repertoire should fill all of the observation time. Thus an observer developing a list of defined behaviors needs to take care that there is not overlap between them. This may be difficult for three reasons:

1. *Atomic level actions or units of behavior may occur as components in more than one molecular or integrated interaction.* It may be worth noting that the human smile may express welcome, pleasure, or aggressive intent even though the same set of atomic components are employed. And many of the same atomic components are employed in other facial expressions such as tense mouth face and fear grimace. Context and interpretation are *not* part of a definition.

2. *Behavior units can often be grouped together and arranged in a hierarchy, and confusion over the level in a hierarchy at which a unit belongs can establish an apparent overlap condition.* As a simple example the observer may note that grooming behavior is a favored activity in most primate species, but that the activity of picking through hair and over skin, extracting debris and parasites, occurs in three possible modes. These are: self- or "auto" grooming where the action is directed toward the subject's own body areas; "giving groom" or one form of "allogrooming" in which the subject actively grooms a different individual; and "receive groom," the other form of allogrooming in which the subject receives the grooming attentions

of another individual. Thus a hierarchy of grooming exists with first-level subdivisions of auto and allogrooming and a second level under allogroom separating giving and receiving grooming. In the parlance of Delgado and Delgado (1962) this would be a "grooming role" divided into active—"giving groom"—and passive—"receiving groom."

3. *A confusion between static and dynamic behavior might exist.* There is a classic beginning observer problem in classification. With the injunction to record a behavior for all of the time involved, students often feel obliged to record whether the subject is "sitting, lying, hanging, prone or supine," and so forth. This then becomes translated into a "behavioral category" and included within the ethogram. While the distinction provided by Delgado and Delgado between static and dynamic initially seems to suggest this is appropriate, an observer must note that these two categories are on different arms of the hierarchy, and hence they should *not* be merged. The reason behind this is that static and dynamic categories imply variables that are different yet can be recorded simultaneously. And they may not correlate with each other. It is entirely possible to record an animal in a particular posture (e.g. sitting) and note that for some of the time it is "resting," for some it is "scanning," and for some it may "receive grooming." The situation is one in which there may be one posture but there are three distinct behaviors being performed. Similarly, while receiving grooming, a subject may be seen as "sitting," "lying supine," "lying prone," and "standing quadrupedally." There is sufficient justification for separation of static and dynamic behavior. However, it should also be noted that the *action of assuming a posture*, such as "sitting up" or "sitting down," *is* an appropriate dynamic behavior and should be recorded as such.

Fixed and Variable Behavior Patterns

While most of the twentieth century has seen a great deal of argument about the causal origin of behavior, the currently accepted paradigm is that most behavior units have roots in both nature and nurture. This is largely a consequence of the fusion between European classical ethology and American comparative psychology, which occurred at the end of the 1960s and led to the defining of behavioral ecology. In spite of this, there are a number of behaviors that can be categorized as being very strongly biologically determined—that is, they belong on the nature side of the issue.

There are two degrees of these natural behaviors: those that are ritualized and those that constitute displays. Behavior is said to be ritualized in an organism when it presents three characteristics: (1) it is stereotypical within the species (it is of the same form and pattern throughout the studied representatives of the species); (2) it can be shown to have been shaped by natural selection; and (3) it can be shown to be strongly controlled by genetic mechanisms. It can be a substantial problem to verify this attribution of ritualized behavior without a very extensive and complex research endeavor; consequently, most of the claims of ritualization are founded upon logical argument and patterns of stereotypy.

The extreme form of ritualization is the display behavior. A display is a ritualized signal that is exaggerated, more stereotypic, and incorporates several elements that make it more complex in form. Almost any behavioral element can be incorporated into a display. A display may include autonomic responses of the sympathetic and parasympathetic nervous systems, such as piloerection (erection of the hair over some area or the entire body), vascular changes (such as color changes in the face, hands, or sexual skin areas due to increased blood flow), intention movements, displacement movements, and other signals. An example of a sequence of display behaviors occurs in the aggressive behavior of most cercopithecoids. The sequence of *headbob*, a brief dropping and raising of the head; *forebob*, a brief dropping of the whole front of the body, looking somewhat like a half push-up; *bounce*, in which the aggressive animal jumps a half step forward, landing on all four limbs at the same time, then jumps back; and *branchshake*, in which either a vertical or a horizontal structure is forcefully slammed by the hind feet while holding on with the forelimbs, often causing the support to sway dramatically, represent an increasing intensity of aggressive signaling. However, perhaps the most common display in the cercopithecoids is "lid," an aggressive signal in which the eyes are blinked to show the contrasting-colored eyelid patches as shown in Figure 3.3.

With these concepts in hand, a student of behavior can proceed to the development of a listing of observed behaviors and can set up appropriate non-redundant, mutually exclusive definitions of the categories.

Figure 3.3 Lid display of aged male Japanese snow monkey (*Macaca fuscata*). A postcard from Arashiyama Mountain, Kyoto (courtesy of M. Nakamichi)

A BEHAVIORAL GRAMMAR FOR DATA RECORDING

As might be evident by this point, a simple list of behaviors is not the only item necessary for the conduct of scientific observation. Also required is a "grammar" for the record construction. There are only a few possible grammars that fill the needs of the observer. The grammar must be simple to learn, straightforward to record, and must adequately represent the stream of behavior of the subject. The most appropriate for general use is one that parallels the subject–verb–predicate structure of most Western languages, in the form of "actor–act–recipient."

The actor–act–recipient grammar is suited to all of the sampling procedures recommended by Jeanne Altmann (1974) and is amenable to utilization in analysis and statistical testing. Because the actor in this grammar may or may not be the focal subject, it is necessary to develop a second set of categories to allow for the focal to passively receive acts from an actor that is not the focal. Since in focal sampling the intent is to record everything that the focal does as actor, the observer must also be equipped to record everything that happens to the focal as recipient.

Our sample record from chapter 2 reflects this grammar in the structure of the sample:

Record of Observation

1*	10:32:22	Jo-Jo receives grooming from Jay-Jay
2	10:34:36	Jo-Jo sits up & scans
3	10:34:37	Jo-Jo locomotes to food dish
4	10:34:41	Jo-Jo eating
5	10:36:12	Jo-Jo drinks
6	10:36:12	Jo-Jo scratches
7	10:36:13	Jo-Jo chases Spike
8	10:36:55	Jo-Jo receives threat from Alice
9	10:36:56	Jo-Jo presents to Alice
10	10:36:57	Jo-Jo receives grooming from Alice

**line numbers added for referencing*

A moment of inspection will result in noting that record lines 2, 3, 4, 5, 6, 7, and 9 are in the active mode—Jo-Jo is *doing* something—while record lines 1, 8, and 10 are in the passive mode—Jo-Jo is ***having something done*** to him. Naturally, if this was a focal animal sample record, then Jo-Jo's name would not be present on every record line but would be in the heading information and a check sheet with this record would look like something like the following:

Focal Subject: Jo-Jo Sample Duration: 10 minutes
Weather: Bright Sun, No wind, Temperature 21°C.
Location: Quadrant B12, small clearing.

10:32:22	receives grooming from Jay-Jay
10:34:36	sits up & scans
10:34:37	locomotes to food dish

10:34:41	eating
10:36:12	drinks
10:36:12	scratches
10:36:13	chases Spike
10:36:55	receives threat from Alice
10:36:56	presents to Alice
10:36:57	receives grooming from Alice

Now that a grammar has been specified (alternative grammars will not be considered in this workbook) and we have some idea of how it contributes to the study of animal behavior, it is appropriate to turn to the practicalities. Writing the data down in long hand, or even typing it out, can be both tedious and prone to error, so a common feature of most recording protocols is a code. Codes can vary greatly between individual observers, just as do the decisions about what the boundaries between behaviors are. In general, a code that consists of one or two or three letters, a *mnemonic*, based upon the behavior label or name is satisfactory. Thus one can set up a codebook in which the behavior label or name is listed along with the code to be used and the definition of the behavior itself. A two- or three-letter code may have the following—"rg" or "rgm" for "receive groom from," "st" for "sits up," "sc" for "scans," "fd" for "feeding/eating," "dr" or "drk" for "drinking," "scr" for "scratching," "ggm" for "giving grooming," and "pr" for "present to"—and would result in the previous check sheet being transformed into a rather more compact record.

Focal Subject: Jo-Jo Sample Duration: 10 minutes
Weather: Bright Sun, No wind, Temperature 21°C.
Location: Quadrant B12, small clearing.

10:32:22	rg Jay-Jay
10:34:36	st/sc
10:34:37	loc dish
10:34:41	fd
10:36:12	dr
10:36:12	scr
10:36:13	ch Spike
10:36:55	ragr Alice
10:36:56	pr Alice
10:36:57	rg Alice

The key feature of the coding scheme is the application of the KISS principle. KISS is a well worn acronym for "Keep It Simple, Stupid," and along with the well-known GIGO rule (Garbage In, Garbage Out), should be applied to the development of a code. Overly complex or unusual mnemonics can make the code confusing to the operator and lead to errors of commission through recording an inappropriate code unit. This can be especially important where the observer is not the originator of the code (i.e., where research assistants are employed by a principal investigator).

The only remaining problem for the observer is accurate time control, and how precisely the time must be recorded. While various timing algorithms have been published for use in computer based data collection programs emphasizing intervals accurate to 1/10th, 1/100th, and even 1/1000th of a second, I fail to see the utility of such precision for field data. In the DATAC 6 program we (Paterson, Kubicek, and Tillekeratne, 1994) made a decision based upon our understanding of the "reaction time loop" of the observer, and set the program parameters to record an accuracy of one (1) second. Our reasoning was this, and it applies equally to the pencil-and-paper recording system: The reaction time loop is the period in which the observer sees the change of behavior, recognizes what the newly initiated behavior is, decides on the appropriate code, and performs the physical actions necessary to make the record—whether writing two to ten characters or hitting them on a keyboard. Under normative conditions, an observer will rarely operate faster than a one-half-second interval, and to be fair to the subject, will an error of ± 0.5 seconds make a significant difference to the data collected? Indeed, over a lengthy study period, the randomization of such observer errors will tend to a net effect of zero. It can be argued in contradiction that human observers and animal subjects can in fact perform behaviors at faster rates, and that millisecond timing is required. While there may be some merit to the argument, the only truly accurate way to deal with such timing concerns is through the use of video or film and slow speed analysis. Such precision is not possible in vivo. It would be unlikely for any observer to attempt millisecond precision timing on any field project; the demands in time and energy, and the resultant volume of data, are excessive.

Operationalizing the Observational Study

Utilizing the KISS and GIGO principles in conjunction with the model of the scientific method, we are now ready to begin the operational planning of the study project. We can proceed to do so using the following outline.

- Do we have a research question? This is an absolute necessity since we will be attempting to find out something about the world.
- Do we have a set of hypotheses? Again, these descend from the research question and must be phrased in a testable fashion (they must be answerable as either yes or no).
- Have we chosen a sampling method? The choice of sampling method depends upon whether the research project is behavior or individual orientated, and is correlated with the recording rule chosen.
- Has a recording rule been selected? Either continuous or interval recording is statistically appropriate, but care must be taken to choose the most relevant method.
- Are the variables set up? Are the variables able to measure the relevant behaviors in useful ways? These form the central core of the research project, without them, and without their verification, the entire project is invalid.

- Have we selected the appropriate statistical procedures for our hypotheses and our variables? It is always a good strategy to have a clear understanding of the statistical procedures relevant to the kind of data to be collected. In many cases the choice of statistic will limit the kind of data and the method of collection.
- Have we dealt with all of the necessary protocol approvals from animal care committees, grant agencies, and governments? The latter are only relevant when a field expedition is required, but the paperwork is part of the process and ACC approval is ***always*** required.
- Have we dealt with the logistics of the project? Health, accommodation, travel documents, customs registrations, biological sample permits, financial transfers, etc.? If our project takes us into the tropics, as is likely for primatological research, we must devote a substantial amount of time to dealing with these issues.

When all of these points have been dealt with, the only things remaining are to actually do the work: perform the observations, collect the data, analyze them, and write up the results. At that point, and only at that point, is a cycle of the scientific process complete, but assuredly the effort has led to the generation of more new questions than have been answered. Welcome to the world of science!

4

A Statistical Primer

"There are lies, damned lies, and *then* there are statistics!" This famous quote, sometimes attributed to Mark Twain, is the byword of many students, but it need not be so. Basic statistical concepts are among the most favored skills of modern society, but the practical application of statistics is something that is not always taught. This chapter is intended to demystify statistics and to present the student with the practical utility of a basic statistical understanding to the completion of the observational and field exercises.

Statistics can be thought of as being an elaborate counting system, and counting items is basic and central to any use of statistical procedures. Statistics are also normally divided into two large categories: descriptive and inferential. To dispose of inferential statistics, since they will not normally be used in any of the exercises, we can note that the central concept of these procedures is to infer from (or on the basis of) a sample, what the conditions of a larger population or "universe" are likely to be. These are the kind of statements we are used to hearing from media reports of surveys—"results accurate to plus or minus 3 percent nineteen times out of twenty." These are probabilistic statements that are basically stating that we think the results we have found from this sampling are accurate to within a range of plus and minus three points and would be so in nineteen samples if we repeated the process on twenty samples. Inferential statistics thus are used to infer or make predictions about other samples, but these tests can also be used to make a mathematical assessment of the validity of the data collected. For the purposes of this workbook, only the field of descriptive statistics need be explored.

Descriptive statistics do exactly what the label applied to them suggests, they describe the characteristics and features of the dataset. As one statistician emphatically stated to me, "descriptive stats ARE your results." They should not be dismissed as "only" the descriptive stats while hurrying on to the more sophisticated inferential analyses. They should be carefully explored and considered, for an extensive mass of information can be very simply extracted through their use.

Our coverage of descriptive statistics will not be like a standard text, it will look at two main categories—measures of distribution and simple comparative statistics—without any attempt at developing or proving relevant theories, and will then deal with the issues of data table construction and the use of Excel or Works programs to mechanize the process. Coverage will proceed with a sequence of terms, some discussion of them, and example calculations based on a simple dataset.

Basic Terms

Two of the most important terms related to statistical analysis of behavior are actually those that divide the data into two kinds: state and event (see chapter 2). The kind of data collected dictates how the statistical description will be done and what kinds of measures can be employed. The term "bout" is often used rather informally in conjunction with these categories even though it has specific definitions, but the two definitions need to be clearly indicated in the context of their use, and bear repeating.

event: Instantaneous or momentary behavior that occurs without measurable duration. The onset of any behavior may also be considered as an event.

state: Behavior with a measurable duration (durational behavior), but may refer to any behavior at a given instant in time, as in the case of scan sampling.

bout: Most often used to refer to a sequence of related durational behaviors, as the interchange of grooming between two partners (where A grooms B five times, B grooms A three times, all occurring within a short period without other behaviors intervening) may be called a "grooming bout." It may also be used to refer to the frequency of durational behaviors; for example, "12 grooming bouts occurred during the 6 hour period."

An alternative when using short interval scans involves counting the strings of observations, a bout consisting of a sequence of uninterrupted same-behavior records.

That is, if the record looks like:

1-grm,
2-grm,
3-feed,
4-grm,
5-grm,
6-feed

There would be *two bouts of grooming and two of feeding*, in spite of there being four records of grooming. The assumption when the intervals between observations are short is that the behavior has actually been continuous and thus it constitutes a single behavioral string.

The terms duration, transition time, and frequency are often related to the differentiation of event from state data, in that event data is considered to have no duration, and it is general practice to ignore transition times. These may be of importance when data are being extracted from video or acoustic records with a fine-grained analytical technique such as frame-by-frame analysis. Frequency is applicable to both event and state data, but is without a time factor associated. It can also be labeled as raw count.

duration: The measurable time over which a behavior takes place.

transition time: The finite period taken by a subject to change from one state to another, or alternatively the onset or termination of a behavior.

frequency: The number of occurrences of a behavior—either events or states may be counted. Note that NO time period is involved. This is also known as the raw count.

The Example Dataset

A small sample of data is utilized in the following series of statistical operations to provide examples of how the calculations may be employed and interpreted. Two behavioral patterns are included from a study of olive baboons: sexual mounting and roar-grunt vocalizations. The definitions of these are not relevant to their use as examples, nor should this data be cited as valid.

- Sexual mounting is observed 100 times in a study of 400 minutes' duration divided into fifteen-minute sampling units or periods. The total duration of this behavior was 123 minutes and the group had twelve members.
- Roar-grunts were recorded 146 times in a total sample of 3,794 behavioral occurrences, and accounted for 378 minutes out of the 23,459 minutes in the study.

Measures of Distribution and Central Tendencies

Two issues are always relevant in any statistical analysis: how the data are distributed and the clear central tendencies within the dataset. These statistics outline the distributional characteristics.

range: The highest and lowest scores of any measure. Typically the range reflects the outside limits of the data.

mean or average The *arithmetic mean* is normally implied, and it is produced by summing all of the individual observations and dividing by the number of observations. Means are calculated as part of the standard deviation example below.

mode: The most common value of the variable in a frequency distribution. It is represented by the greatest number of individuals in a grade distribution, for example. Distributions that display two peaks are "bimodal," and those with more than two peaks are "multimodal."

median: The value of the variable in an ordered array or sequence that has an equal number of items on either side of it. Thus, it represents the midpoint of a distribution.

These three statistics may coincide at the same value on a gaussian or bell-shaped curve; most often this case suggests the curve represents a "normal" distribution.

standard deviation: The standard deviation (SD) is the most commonly used measure of dispersion, that is it measures the degree to which values are spread away from the mean value of a distribution. It is defined as the square root of the squared deviations of the scores around the mean, divided by the number of scores.

The formula for the standard deviation is:

$$\mathrm{SD} = \sqrt{\frac{\Sigma d^2}{N-1}}$$

Where ***d*** is the deviation of an individual value from the mean, Σ (sigma) is the symbol for "sum of ," and ***N*** is the number of cases. The formula can be read as: "The standard deviation of the sample equals the square root of the value yielded by the sum of squares of the deviations divided by the number of cases minus one." The "sum of squares of the deviations of the measurements from the mean measurement" is necessary to this calculation and the formula for it is:

$$\Sigma d^2 = \Sigma x^2 - \frac{(\Sigma x)^2}{N}$$

Where ***x*** is each measurement value. There is, however, a concise and practical way to calculate these statistics.

You will need to obtain the following:

- The sum of the measurements (i.e., add up all of the individual measurement values, the *x*s, this is Σx).
- The sum of the squares of the measurements (i.e., square each measurement and then add up all those values = Σx^2).
- The square of the sum of the measurements = $(\Sigma x)^2$.

- The sum of squares of the deviations of the measurements from the mean, the last formula given above, = Σd^2.

An appropriate format for calculating the standard error of measurement is a table setup as follows (NOTE: Do *not* round off any values until the *end* of the calculation; rounding at any earlier stage will invalidate the calculation):

Measurement Number	**Measurement (x-values)**	**x^2 values**
1	10.000	100
2	10.016	100.320256
3	9.997	99.940009
4	9.991	99.820081
5	10.023	100.460529
6	10.002	100.040004
7	9.998	99.960004
8	9.996	99.920016
9	10.011	100.220121
10	10.000	100

Add each column to get: Σx Σx^2

- Sum of measurements $\Sigma x = 100.034$
- Sum of squares of the measurements $\Sigma x^2 = 1000.681020$
- Square of the sum of the measurements $(\Sigma x)^2 = 10006.80116$
- Sum of squares of the deviations of the measurements from the mean

Insert the arithmetic values

$$\Sigma d^2 = \Sigma x^2 - \frac{(\Sigma x)^2}{N}$$

$$\Sigma d^2 = 1000.681020 - \left(\frac{10006.80116}{10}\right)$$

$= 1000.681020 - (1000.680116)$

$= 0.000904$

Therefore the standard deviation.

$$= \sqrt{\frac{0.000904}{9}}$$

$= \sqrt{0.0001004444}$ *(recurring)*

$= .01002219537$

rounded to SD = ± 0.01

Frequencies and Rates

Some of the most basic calcuations are founded upon the frequency data nd correlate the raw count with the time function, or in relationship to some other related count.

relative frequency: This indicates the probability of a particular behavior being observed at a randomly selected behavior change, or the probability of one particular behavior of the known set occurring.

Relative frequency (A) for behavior changes:

$$\frac{\text{Frequency of one behavior}}{\text{Total number of behavior changes}} = \text{relative frequency } A.$$

This gives the relative probability of a behavior change.

Relative frequency (B) of defined behaviors:

$$\frac{\text{Frequency of one behavior}}{\text{Total number of identified behaviors}} = \text{relative frequency } B.$$

This yields the probability of the behavior occurring as a given behavior at a particular time.

Example: In a particular sampling, the behavior of grooming another group member occurs 36 times in a total sample of 479 behavior changes (transitions from one behavior to another), thus the relative frequency A or probability of change is

$$36/479 = .075.$$

Since there were only 65 different (defined) behaviors seen, the relative frequency B or probability of grooming occurring as the behavior at a particular sampling time is

$$36/65 = .554.$$

Interpretation is straightforward: in the sample recorded, the probability of any behavior change involving a switch to "grooming" is quite low (0.075), but the probability of grooming being the behavior at any particular sample time is quite high (0.554). A is a relative frequency of occurrence, while B is a relative frequency of the behavior set.

rate: The frequency per unit of time, thus requires that the duration of the sample be known. Having equal sample periods makes the calculation easier but the calculation is still valid with unequal durations if the total observation time is divided into the sum of the frequencies for all samples. Rates are most useful when a common time base such as "frequency per hour or per minute" is employed. Thus there are two variants of the calculation.

Frequency per sample unit:

$$\frac{\text{Number of occurrences}}{\text{Number of sample units}} = \text{“rate” (frequency per sample unit)}$$

Frequency per time unit:

$$\frac{\text{Number of occurrences}}{\text{Total observation time}} = \text{“rate” (frequency per hour or minute)}$$

In any presentation of a rate, it is *obligatory* to present the units involved.

Example: Sexual mounting is observed 100 times in a study of 400 minutes' duration divided into fifteen-minute sampling units (400 minutes equals 26.67 sample units). The rate of sexual mounting is thus:

100/400 = 0.25 *mounts per minute,*

or 100/26.67 = 3.75 *mounts per sample period.*

hourly rate: This is variation of the rate calculation using "hours" as the time unit.

$$\text{Hourly rate} = \frac{\text{frequency of behavior}}{\text{hours of observation}}$$

Example: The sexual mounting data would yield an hourly rate of

100/6.67 hours = 14.992 *times per hour.*

Since the number of significant digits appropriate here is zero, the rate would be rounded to fifteen per hour.

Probability, Proportion, and Percentage

These three terms are all closely related and very similar to the calculations for relative frequencies, though the context and use of these are different.

proportion: The decimal expression of any fraction. The formula appears to be essentially the same as for relative frequency, but the formula is also used with durations (time):

$$\frac{\text{Behavior } A \text{ frequency (or duration)}}{T \text{ total frequency of all behaviors (or durations)}} = \text{Proportion } A \text{ of } T$$

Example: Roar-grunts in a baboon field study were recorded 146 times in a total sample of 3,794 behavioral occurrences and continued for 378 minutes out of the 23,459 minutes. Therefore the proportion or relative frequency of roar-grunting was 146/3794 = .0384, and the proportion of time was 378/23459 = .0161. Proportion can be viewed as the generalized condition, while the two relative frequency calculations are specific subcategories of the model.

probability: The proportional expression of the likelihood of a particular behavior occurring. This occurs on a scale of zero to one, with the maximum probability of 1.0, meaning that the behavior *always* take place, while a probability of 0.0 (the minimum) means that it *never* occurs.

Example: In the baboon dataset, the probability of hearing roar-grunts during any particular minute of the study is 0.0161. But a simpler, and more direct, example is: If a study shows that during the hour from 9 to 10 o'clock 76 out of 100 students are seated at their desks, the probability of any one student being seated during that hour is 76/100 = .76.

percent: The same calculation as is done for proportion but multiplied by 100. That is, a .76 proportion equals 76%.

Comparative Statistics

The following calculations are a bit more complex than the preceding have been, but they have a much greater utility in that they can be used to compare between groups and individuals. These calculations are thus inherently useful for the description and understanding of the variations between individual primates and between larger aggregations such as groups or species.

mean duration per bout (MDB): Total duration of a behavioral category divided by its frequency. This provides a measure of how long, on average, each behavior continues once it has begun.

$$\text{mean duration of behavior } A = \frac{\text{Sum of durations of all } A}{\text{Total frequency of } A}$$

Example: In the baboon sexual mounting data, the total duration of this behavior was 123 minutes. Therefore the mean duration of sexual mounting was 123/100 = 1.23 minutes (each mount takes an average of 1.23 minutes).

mean duration per hour (MDH): The average number of minutes per hour in a state equals the total duration in minutes divided by the number of hours of observation. Note that the resulting units are "minutes per hour".

$$\text{mean duration per hour of } A = \frac{\text{Sum of durations of } A \text{ in minutes}}{\text{Total hours of observation}}$$

Example: For the baboon sexual mount data, the mean duration per hour would be 123/6.67 = 18.44 minutes per hour.

Mean duration per hour for interval sampling: When interval sampling is used, the difficulty lies in the calculation of the duration of the behaviors, and the not very precise relationship with real time. If a study results in 12,000 samples taken 10 seconds apart, the time involved is 12,000 x 10 = 120,000 seconds = 2,000 minutes = 33.33 (recurring) hours. If a particular behavior is scored on 1,289 samples, and we make the obvious assumption that the record represents 10 seconds (the maximum possible) of that behavior, then the duration is 1,289 x 10 = 12,890 seconds = 214.83 minutes = 3.58 hours. This behavior's mean duration per hour then is 214.83/33.33 = 6.446 minutes per hour.

mean rate per individual (MRI): Total frequency for all individuals divided by the total observation time for all individuals all divided by the number of individuals in the study group. The resulting units thus become "occurrences per minute per individual." If the observation time is entered in hours, the units become "occurrences per hour per individual."

$$\text{mean rate of } A \text{ per indiv.} = \frac{\dfrac{\text{Total frequency of } A \text{ for all indiv.}}{\text{Total observ. time (hours) for all indiv.}}}{\text{Number of individuals}}$$

Example: For the sexual mount data, since the group has twelve members, the mean rate per individual is 100/400 = .25 mounts per minute divided by 12 = 0.0208 mounts per minute per individual, or 0.000346 mounts per hour per individual.

mean duration per individual (MDI): Total duration for all individuals in minutes, divided by the total observation time for all individuals in hours, all divided by the number of individuals in the group. Resulting units for MDI are "minutes per hour per individual."

$$\text{mean dur. of } A \text{ per indiv.} = \frac{\dfrac{\text{Total duration (minutes) of A for all indiv.}}{\text{Total observ. time (hours) for all indiv.}}}{\text{Number of individuals}}$$

Example: For the sexual mount data, the group has twelve members, and the mean duration per individual is 123/6.67 = 18.440779 minutes per hour divided by 12 = 1.54 minutes per hour per individual.

An alternative mode of calculation for MRI and MDI, but only if the sample durations are equal for all subjects, involves using the individual mean rates (calculated as total frequency or duration for individual divided by the observation time for the individual). These can be summed and divided by the number of individual subjects.

Percentages of Time for Continuous and Interval Sampling Protocols

Amongst the most common calculations in primatology are those that attempt to show how the subjects allocate their time, especially when some comparison between populations, environments, or species is being undertaken. Because continuous and interval sampling produce distinctly different kinds of data, caution must be employed in selecting the correct calculation model.

percent of time for continuous sampling: Often it is appropriate to calculate percentages of time as a preliminary to producing comparative tables or graphs of behaviors. For each behavior the calculation would be:

$$= \frac{\text{mean duration per hour}}{\text{60 minutes}} \times 100$$

Example: For the sexual mounting data, where 123 minutes of 400 (6.67 hours) were observed in this activity, the percentage time would be (123/6.67)/60 x 100 = 30.73% of the time was taken up in sexual mounting.

percent of time for scan sampling: It is often desired to calculate the percentage of "time" when using an interval sampling protocol; however, the result is actually a percentage of scans and is only an approximation of the percentage of real time. This technique has been carefully investigated and the error rates for different intervals suggest that it may be used only when the scans are less than ten seconds apart if the calculation is expected to be valid in any degree.

$$= \frac{\text{number of scans with behavior scored}}{\text{total number of scans}} \times 100$$

Example: In a scan sample study, 12,000 scans were taken at ten-second intervals, in 8,675 scans "resting" was recorded, and in 2,347 scans "locomotion" was recorded. The percent of time resting is 8675/12000 x 100 = 72.29%, and the percent of time locomoting is 2347/12000 x 100 = 19.56%. The caveat is that these are actually a percent of scans and not directly measured time.

STATISTICAL REPORTING

In any report that employs statistical description of results, some tabular features are always appropriate and expected. The two main items are tables dealing with the frequency and duration data.

- *Frequency table*. This should show the raw frequency and the rates for the behaviors being reported. An example of a basic frequency table follows as Table 4.1.
- *Duration table*. This should show the durations of behaviors being reported unless the study employs interval samples with more than ten seconds between records, in which case no duration table is expected.

Note that one additional row and column—called the "marginals"—are *necessary* components of every table. They provide totals for each row and each column. When constructing a table, the marginals are an important safety check. In adding up the total row or the total column, the number produced must be the same in each direction—otherwise there is an error within the data presented. The data in the table also allows some interpretation simply through inspection. In Table 4.1 it should be evident that the juvenile engages in more frequent behavior than the adults, and that the female is more than twice as active as the male. It also shows that while play behavior may be the most frequent activity, virtually all of it is done by the juvenile, and similarly while grooming is prominent, the female does nearly all, and the male does none. Such information can be directly interpreted and differences between age or sex classes and between different behaviors can be easily seen and presented.

Table 4.1 Basic frequency for a study involving three age-sex classes

Behavior	Ad-Male	Ad-Female	Juvenile	Row Totals
Resting	54	32	8	94
Feeding	34	61	41	136
Locomotion	12	19	98	129
Grooming	0	118	7	125
Playing	3	0	135	138
Column totals	103	230	289	622

A slightly more complex frequency table would look like Table 4.2 with the presentation of both frequency and rate data. A significant difference exists in this table with respect to the information in the marginals: It is optional to add up the rates in the column totals row, and the row total should not be added at all. The reason for proceeding in this way should be obvious: If column totals for rates are in place, as with the bracketed values in Table 4.2, the only thing that they demonstrate is the overall rate of behavior for the category. It is also important to note that the rates in the Row Totals column are NOT sums from the rates in the other columns; they are *calculated rates* from the total frequency, which actually is a sum. If the exercise of adding the rate values for the marginals is undertaken, it will not be equal (sum of column totals = 124.4, sum of row totals = 41.47). One further feature of Table 4.2 is the presence of abbreviations in the table and the consequent necessity for a key or legend to explain them and present the units employed.

Table 4.2 An example of a basic frequency/rate table

Behavior	Ad-Male		Ad-Female		Juvenile		Totals	
	f	r	f	r	f	r	f	r
Resting	54	10.8	32	6.4	8	1.6	94	6.27
Feeding	34	6.8	61	12.2	41	8.2	136	9.07
Locomotion	12	2.4	19	3.8	98	19.6	129	8.6
Grooming	0	0.0	118	23.6	7	1.4	125	8.33
Playing	3	.6	0	0.0	135	27.0	138	9.2
Column Totals	103	[20.6]	230	[46.0]	289	[57.8]	622	

f = raw count, r = rate in occurrences per hour.
(5 hours observation on each subject)

Duration tables are similar in construction but hold different data. Table 4.3 presents some duration data and relevant calculations performed with the data. The data available for Table 4.3 allows the calculation of MDH for the three cases of the adult male, the adult female, and the totals, but note once again that calculations such as MDH, MDI, and percent of time *may not be summed* from the body of the table, nor will they yield the same values for row and column totals. The addition of the MDI calculation provides a value that would be of use in comparative examination of the differences between this and another group. It should also be noted that if MDI were calculated for the individual male and female in this case, it would equal the MDH. Such would not be the case if several members of each class were aggregated in the table.

Table 4.3 An example of a basic duration table employing data on one male and one female

Behavior	Ad-Male				Ad-Female				Totals				
	f	dur	MDH	%Time	f	dur	MDH	%Time	f	dur	MDH	MDI	%Time
Rest	54	178	35.6	59.33	32	41	8.2	13.66	86	219	21.9	10.95	36.5
Feed	34	43	8.6	14.33	61	50	10.0	16.66	95	93	9.3	4.65	15.5
Loco	12	57	11.4	19.0	19	12	2.4	4.0	31	69	6.9	3.45	11.5
Groom	0	–	–	–	118	197	39.4	65.66	118	197	19.7	9.85	32.83
Play	3	22	4.4	7.33	0	–	–	–	3	22	2.2	1.1	3.66
Col. total	103	300	60	99.99	230	300	60	99.98	333	600	60	30	99.99

5 hours (300 minutes) observation on each individual. f = raw count, dur = duration in minutes, MDH = mean duration/hour in min/hour, MDI = mean duration per individual in min/hour/indiv.

Spreadsheet programs such as Excel, or the spreadsheet components of omnibus programs such as ClarisWorks or Microsoft Works, can be set up in a relatively simple fashion to construct tables and perform the relevant calculations. Most word processing programs like Word can construct tables and even format them for presentations. A "works" or Excel program is a useful adjunct to a straightforward word processor and may have the capacity to link results directly to a table in the final text. Developing capability in connecting material between text and computational documents is a worthwhile skill for more than the primatology course.

5

The Preparation of Scientific Reports
A Beginner's Guide

Expectations for a Student Report

Scientific reporting varies a great deal. Reports may be very terse, full of jargon, and essentially incomprehensible except to another specialist in the particular subfield, or they may be exceptionally fine literature. At either extreme, however, they will tend to follow the generalized format detailed in the following pages. For a student involved with the study of primates, the expectations are somewhat similar in that what is expected is a literate essay organized in the scientific report format, with sufficient tables, graphs, and/or figures to illustrate the text. It is expected that a student will need to conduct some examination of library resources in addition to performing observations, and must cite and reference the materials used. For this latter, the citation mode expected in primatology is one of the American Psychological Association (APA) formats simply designated as author-date (page numbers are included only for direct quotes). The pattern may be illustrated as:

As Smith and Jones (1893) found in their study of rabbitoes, the first action is . . .

or

Male baboons weigh between 40 and 55 pounds (18 to 25 kilos) when captured in the wild (R. Coble, personal communication, October 1969; DeVore and Washburn, 1963).

Very few primatology journals make use of footnotes, and the general direction is to avoid them. Every publication cited in the text of a paper or report must make an appearance in the bibliographic listing at the end of the document. It is of little real concern whether that component is labeled References, Sources, Citations, or Bibliography. What matters is the consistent use of the same format throughout, and complete referencing of all cited, borrowed, quoted, or paraphrased materials used in the generation of the report. Several bibliographic programs are available for all major computer platforms to scan a document and generate a full reference listing. These are programs generally able to produce the reference listing in several different formats.

Every report is expected to be presented in competent and correct English. *Thus it is important to proofread the document and correct the spelling and grammar before submission.* Now that computer use has become universal in postsecondary educational systems, and with the general availability of word processing software with both spelling and grammar checking modules, there is no excuse for submitting an illiterate document. Good writing skills can be learned by anyone, and with computer assistance there is little excuse for turning in a poorly prepared document.

Writing for Primatology—Guidelines for the Scientific Report

Scientific reports in primatology follow the more or less standard pattern of: introduction, subjects and methods, results, discussion, and conclusion. In addition there will often be illustrative material in the form of tables, graphs, and figures, perhaps appendices with explanatory material or data lists, and a reference listing.

Contrary to most writers' practice, the introduction should actually be written as the last component of a report, once everything else is complete, and the writer has a clear understanding of the project as a whole. The introduction leads the reader to the problem, theme, and content of the report, and in several ways may be thought of as an extended abstract briefly presenting the problem in the context of the discipline, showing why the problem is a significant one, or is otherwise worthy of investigation, and perhaps summarizing the work that has been done on the topic in the past. It is essential for an introduction to cogently present the reason for undertaking the project, and this does *not* normally include, "because it was required for the course"—that is a truism that need not be repeated. Thus the introduction presents the problem, how it was approached, and what the results of the investigation were. It must encourage the reader to see the importance of the rest of the document and lead the reader smoothly into the next section.

One point that separates a research paper from a thesis is the *literature review*—it is not considered to be a requirement in a research article but is normally part of a master's or doctoral thesis or dissertation. Literature reviews may take different forms but are generally seen as an extensive discussion of the relevant publications linked to the theory and data surrounding the issues that are central to the thesis. They are often typified by long strings of citations of works listed as supporting or denying support for a hypothesis, or providing comparative data about a species, ecosystem, habitat, and so forth. For the purposes of a beginning student of primatology, it is necessary only to be able to recognize the need for and characteristics of "literature review," though it may be important if the instructor decides that one is a necessary component of a report.

The *Subjects* and *Methods* portion is to describe the population or set of materials that were used in the study, and how the study was conducted. This does not necessarily have to be a deadly dull listing such as "thirty-two subjects of *Macaca fuscata* were subjected to training for ten trials followed by two test sessions in a standard Skinner enclosure," but can and should be written in a literate and engaging style. In order to inform the reader adequately of the scientific procedure employed in the study, the report needs to present sufficient details about the subject animals and their habitats and the methods employed in the study. It is important to describe the makeup of the social group, its habitat, special conditions, age and sex composition, as well as provisional status rankings for each group member. This may be conveniently presented as a table. Under methods it is important to specify what kind of observational procedure was employed, what the observing and recording protocols were, and any special rules used by the observer in the handling of the data. Where summarization and statistics (including graphs) are presented, the analytical mechanisms and programs used should be noted. If formal logical hypotheses are to be employed in the study, this is the location for their presentation. The essence of this section is informing the reader what was done, who and where the subjects were, and how it was done. The reader should be able to judge the scientific validity of the study from this section alone.

The *Results* section is often the worst prepared component of a student report. It is NOT sufficient to direct the reader to look at tables "1 through 37" and graphs "1 to 16." While there is no precise statement of what this section must or must not contain, it is important to recognize that a complete listing of the raw data is NOT to be presented. The results section is expected to be a *written* component, pointing out the important features and findings in the data. The results and observations usually will consist of summarizations of the observations, condensed descriptive statistical presentations, and perhaps a few notations of exceptional or unusual observations. It will normally be in this section that the majority of tables, graphs, and figures will appear, and these are expected to be appropriate summaries derived from simple descriptive statistics. Overall, what this section does is present a picture of the results of the study, but discussion of those results is left to the next section.

The *Discussion* section is normally the most important part of the paper, it is here that the consideration of the data and how it relates to the problem or theme takes place. The significance and meaning of the findings presented in the results section is discussed within the context of the methodology of the study and in relation to the previously cited literature. Discussion sections may be extensive, with expanded discussions pointing to the importance of findings shown in the tables and graphs, or may be very short, or even merged into the conclusion section. One useful rule is that the discussion should not be the largest section of the paper (Hailman and Strier, 1997). The overall responsibility of the discussion section is to show that the project fulfilled its objectives, or did not, and to present a considered concise analysis of the results. A discussion will also normally reference the relevant materials in the literature and attempt to reconcile the new findings with the older information. This section is expected to deal with development toward reaching a conclusion relevant to the problem or theme that was presented in the introduction.

The *Conclusion* is also a frequent problem for writers. This can be seen in published as well as student reports—they may range from terse, abbreviated listings of the major conclusions, up to lengthy discourses that do more to confuse the issues than to enlighten them. It is recommended that conclusion sections should be of modest length and, in effect, a concise summing up of the report. This is the place to present conclusions drawn from the results outlined in the earlier sections. As an appropriate length recommendation, one page is normally sufficient for a term paper or article, perhaps five to ten pages for a graduate thesis. The conclusion should deal with all of the questions posed in the introduction, and most critically it MUST be based on the data collected. In other words, the results, followed by the discussion, should lead the reader to already hold the same perceptions as found in the written component. The most important component of the conclusion is a clear answer to the question proposed in the introduction. A classic and extremely embarrassing student *faux pas* is to provide a concluding statement that clearly denies all of the work and data presented in preceding sections. While the evaluator may appreciate humor, this is unlikely to yield rewards.

Illustrative Materials

The common illustrative materials employed in scientific reports are tables, figures, photographs, and graphs. Both tables and graphs can appear in innumerable styles and forms, but each has the basic features that are common to the class. Tables are summaries of data presented in rows and columns that enable comparisons to be made between categories and between individual categories and sum totals. The formats for common frequency and duration tables was discussed in chapter 4. Graphs are typically two-dimensional visual representations of data that could be, and often are, presented in tabular form. There is a certain degree of redundancy between tables and graphs; preference for one or the other depends on perceptual pattern and taste, but tables provide exact data while the graph gives only an approximation. It is, however, rapidly becoming recognized that visualizations of data can convey significantly more information in a short period of time to the reader than can a dense table of numbers.

Each unit of illustrative material—a table, a graph, or a figure—should, indeed must, stand on its own. It should be possible to understand and interpret the data contained within the illustration and its caption without reference to the text body of the document. It is important that the data presented in a graph be easily identified and assimilated, thus it is important to keep graphic presentations as simple and clean as possible. At the same time it is important that each be complete. Each table, graph, photograph, or figure needs a title, appropriate labels, and an explanatory caption. Sometimes photographs are the only appropriate mechanism to illustrate a specific situation, condition, or response pattern. The amount of information present in a photograph is much higher than in any graphic, but it must be very carefully framed, cropped, and captioned to be useful. It is generally inappropriate for a photoessay to be submitted unless specifically asked for; however, a good photograph is never out of place as an illustrative elaboration.

As illustrative material, the following examples of tables, graphs, photos, and figures are presented and their features discussed.

Table 5.1 shows the subject animals, a subset of the adult males of the Arashiyama "A" troop, which was moved to south Texas in 1972, in a study

Table 5.1 Sample males in postural study

Short-bodied Males

Tatoo Number	Common Name	Family	Birth Year
001	Fatboy	Rheus	1971
028	Suma	Suma	1964
064	Moe	Momo	1964
128	Meme65 (Hulk)	Meme	1965
134	Ran68	Ran	1968

Long-bodied Males

Tatoo Number	Common Name	Family	Birth Year
002	Dimbulb 2	Pelka	1970
055	Wania One Eye	Wania	1970
074	Fang	Rotte	1970
129	RockyTwo	Betta	1971
143	Dimbulb 1	Pelka	1968

of their postural behavior. Males selected for this study were separated into long- and short-body length categories. Note that most of the short-bodied subjects were born in the 1960s and most of the long-bodied subjects were born in the 1970s. The table is provided with a title, the columns are labeled, each subgroup is identified and provided with a small illustration symbolizing the differences, and the rows are filled in with the appropriate data. This table shows the tattoo number on each subject's face, the common name used by the colony manager, the matriline to which the subject belongs, and his birth year. At the same time the table clearly shows the two categories into which this population of males has been divided and gives some indication of the criterion used to do so.

While tables are common mechanisms to present substantial masses of data in an organized format, the information often becomes more useful when presented in the form of a graph. And it has become quite simple to generate graphs in a multitude of styles, forms, and variants with readily available computer programs. The data analysis program Excel, on both Macintosh and Intel processor platforms, is one program with the ability to integrate the analysis functions with the capability of generating graphs. However, programs such as Harvard Graphics on Windows computers, and CACricket-Graph III on the Macintosh can generally produce graphs more easily and in greater variation than an integrated analysis package or MSWorks–type program. It will be profitable to invest a few hours in learning to operate a graphics program, but it is also necessary to bear in mind that a graph *requires* a title, values and scales on axes, legends, and keys. A graph, often enclosed within a box on the printed page, *should be interpretable by itself, without the need to refer to the accompanying text* in order to understand the presentation. Figure 5.1 shows the features expected in a bar graph, and Figure 5.2 presents a graph with several overlaid blocks of data. Figure 5.1 shows a title for the graph, labels for both the horizontal and vertical axes, and a legend. The values and tick marks on both axes are indicated by the italic comments. Legends are much more important when more than one set of values is being illustrated in a stacked graph. Note that the bars on the graph are two-dimensional. This is how the bars should appear. It is best to avoid programs that make use of a feature called "add depth" to produce a pseudo three-dimensional effect. This has the perceptual effect on the viewer of seeing the columns as representing volumes, not the area nor the heights of the columns. The latter is the only real expression of the data. In order to avoid giving false impressions, it is best to avoid using these features.

Another misleading representation often occurs when error bars are employed to show the range around a graphed value, commonly a mean. Note that the values of errors or ranges are not necessarily symmetrical and, consequently, employing them on a column or bar graph leads to the generation of a "pinhead" graph, with only part of the errors or ranges showing. If the data is indeed asymmetrical, this type of graph hides it. In general, pinhead graphs are to be avoided. Figure 5.3 correctly shows the use of error or range bars on a line graph.

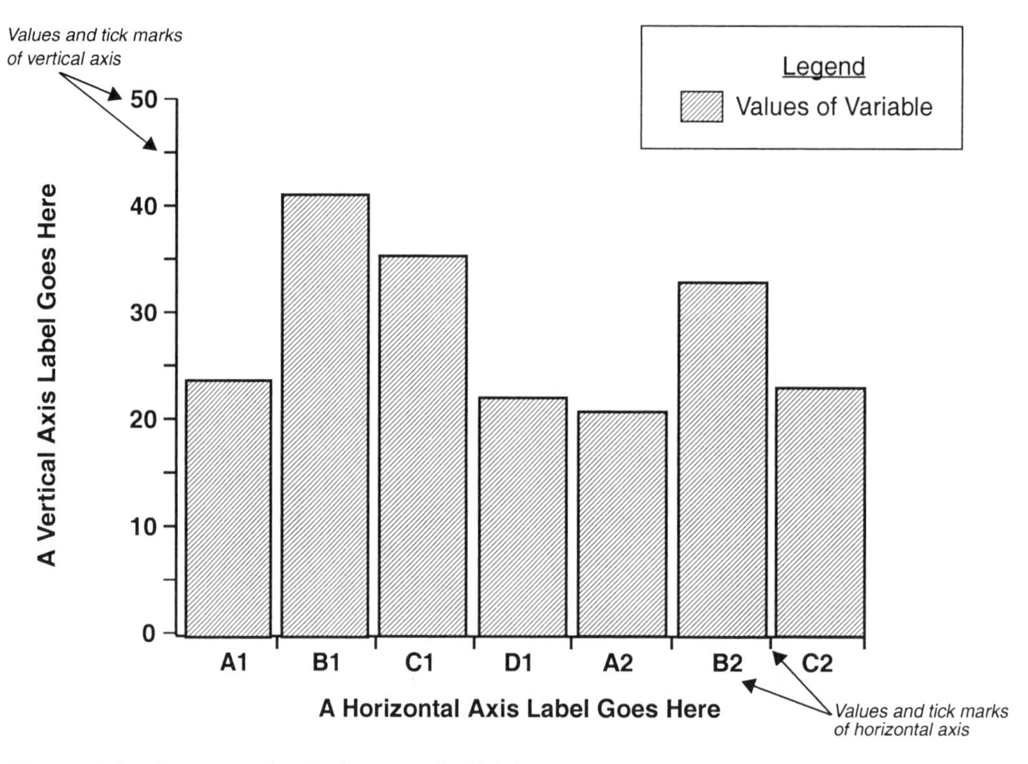

Figure 5.1 An example of a bar graph (title).

Figure 5.4 presents a distinctive and very useful graph, one that represents the differences from the mean (known as the "deviations" and used in chapter 4 for calculation of standard deviation). This type of graph is based on data derived by subtracting the average value for the species behavior from the average for all species (at least in Figure 5.4).

Values derived will be either positive or negative and when entered into a spreadsheet will produce this type of graph. This form is also valuable for showing differences between individuals or age and sex classes within a group, and may be produced in the same way.

Images used within a report are treated similarly: each must have a title, appropriate labels, and appropriate captions as shown in Figure 5.5. As should be obvious by now, it is also customary to number tables and graphics (including images) separately and in sequence.

Photographs, being another form of image, should also be provided with a label, title, and an appropriate caption as demonstrated in Figure 5.6. Modern computer technology, especially the recent reductions in cost of scanning devices, makes it possible to incorporate photographic images in gray scale into reports, with ease. Color imagery, of course, requires the use of a color printer, and the majority of recent ink-jet printers are capable of providing a useable print quality output.

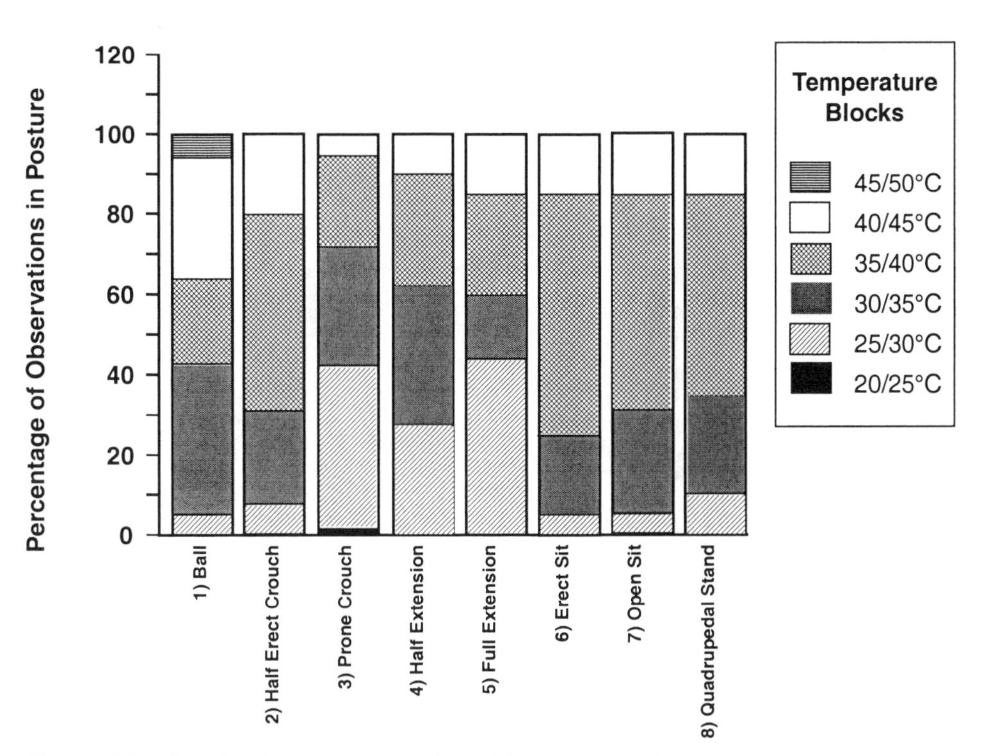

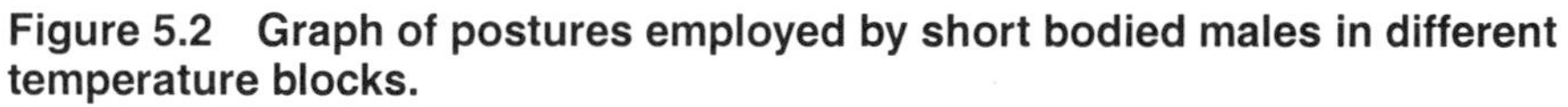

Figure 5.2 Graph of postures employed by short bodied males in different temperature blocks.

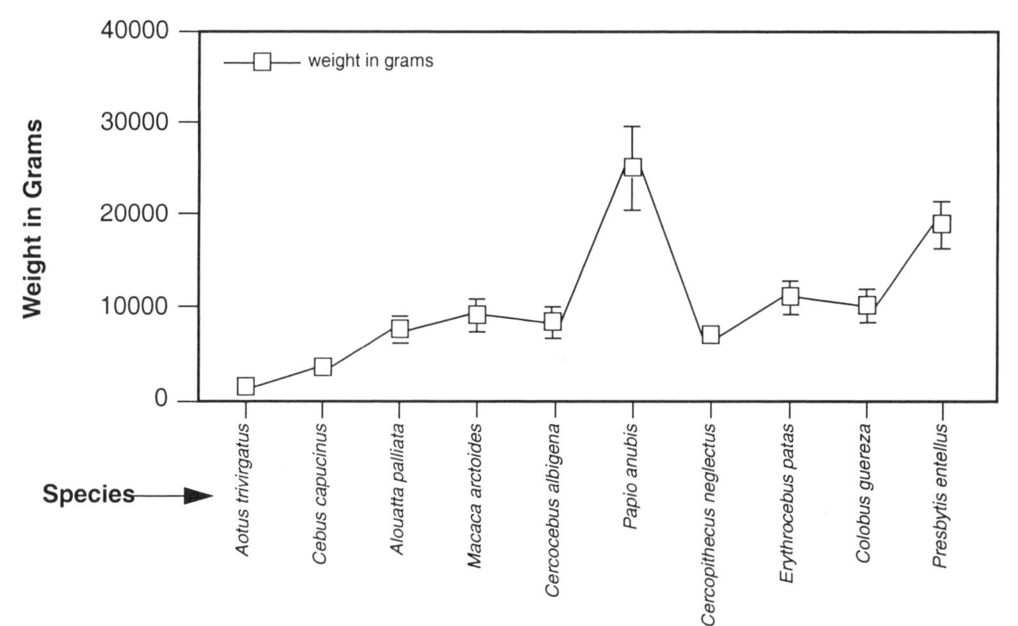

Figure 5.3 Line graph of primate weights with error bars.

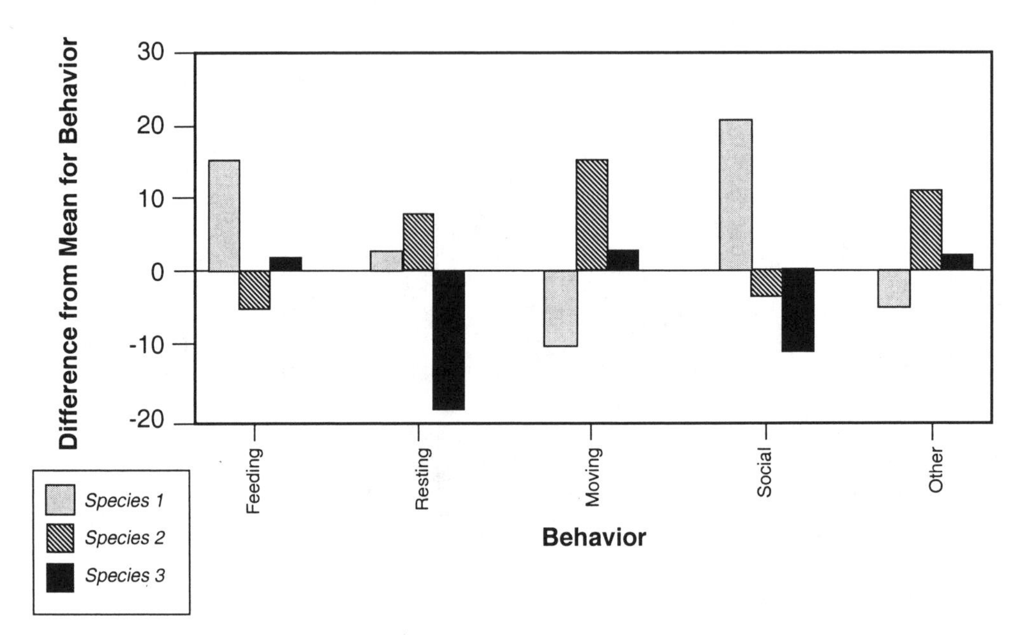

Figure 5.4 Graph of differences from mean values in time budgets of three species.

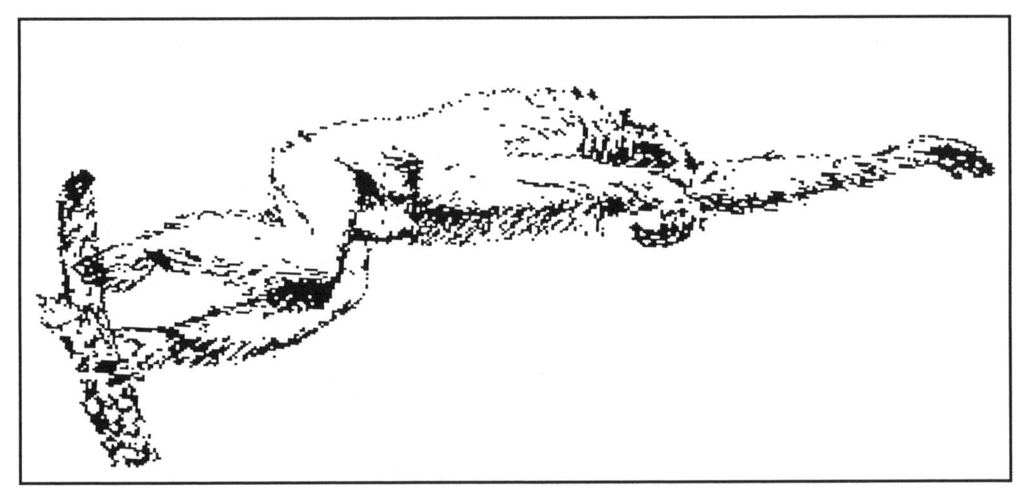

Figure 5.5 Take-off posture of *Hylobates lar.* Digitized image from a drawing in Oxnard (1983) based upon cinematic film footage. One of a series of sequential illustrations of movement dynamics.

Figure 5.6 Masturbation by an adult male olive baboon (*Papio cynocephalus anubis*) in May 1970 at Queen Elizabeth National Park in Uganda. Photo J. D. Paterson.

APPENDICES

Appendices are not normally utilized in a student report, but they occasionally turn up in professional papers and are a common feature of books and monographs. An appendix is used to expand upon a particular aspect of the document and may at times assume the characteristics of an expanded footnote. In other cases, especially where a mathematical proof for an ecological mechanism is needed as supportive evidence, the appropriate location for it is in an appendix. Appendices are also used to present extensive listings of species located in an environment or to provide a complete ethogram in cases where the data is relevant but placing it within the text would disrupt the line of thought being presented. The essence of an appendix is that it provides something extra to the reader but is not necessary to an understanding of the document.

BIBLIOGRAPHIC FORMATS

As noted earlier, the most common citation pattern in primatology is that of the APA author-date format, adding page numbers only for a direct quote of more than three lines; bibliographic references also follow variations on the standard APA format. The references employed in this chapter are set out below. Note the pattern: Journal names (*Behaviour*, *Primates*) should be set in italics, and book titles are treated the same. The difference between journal titles and book titles is that initial capitals are used for all the appropriate words in journal titles and initial caps are used only on the first word, the word immediately following a colon, or proper nouns in book titles. Personal communications are not placed in the bibliography; instructors may, however, wish citations of their class notes or lectures to be included in the bibliography.

Critical Reading and Reviewing in Primatology

A critical approach to the reading of scientific literature is an important aspect of any research work, and it may be the case that a student will be asked to produce a critical review of a set of research papers as a component of a course. It is important to recognize that this process of critical reading is not simply "turning on the bullshit detector," nor is it intended to be a route to trashing or praising the author for style or content alone. Most appropriately a critical review should be directed at the methods and their execution as presented in the paper. A series of questions may serve as a guide in this endeavor.

- What are the research questions posed by the author(s)? Are these clearly presented? Are they understandable?
- What are the formal hypotheses being tested? If they are not visible, how are they implied or are they to remain a mystery?
- What variables have been established to evaluate the hypotheses? How are these variables "measured"?
- Do these variables and their measurements have the logical capability of rejecting the hypothesis? If they do not, is this then a "scientific" inquiry?
- Has the author incorporated all of these aspects into a cohesive whole? Has the author done what was proposed in the research questions? Have the hypotheses been appropriately tested?
- Only at this point should the reviewer be prepared to turn to issues of style, logic, and grammar in the final product.

Any student in primatology is expected to learn to read critically among the primary literatures available. And this brings up the point that one should definitely NOT rely upon "secondary literature" (this comes in two forms: textbooks that synthesize the work of others for general consumption and that may be subject to biases of interpretation and meta-analyses where an author reanalyzes or performs a secondary analysis on data collected by others). Reliance on the latter form is acceptable to a limited extent in discussions of secondary sources, but reliance on the former is not. In relying on secondary sources, students are abdicating responsibility for evaluating and understanding the original work for themselves, leaving it to another individual who may or may not have similar or different perceptions of it. One of the most consistent error patterns seen in beginning primatology students is an excessive and inappropriate reliance upon secondary literature.

BIBLIOGRAPHY

Altmann, J. (1974). Observational study of behavior: Sampling methods. *Behaviour 49,* 227–267.

Baulu, J., and Redmond Jr., D. E. (1978). Some sampling considerations in the quantitation of monkey behavior under field and captive conditions. *Primates 19*(2), 391–400.

DeVore, I., and Washburn, S. L. (1963). Baboon ecology and human evolution. In F. C. Howell and F. Bourliere (Eds.), *African ecology and human evolution* (pp. 335–367). Chicago: Aldine/Atherton. Publishers.

Hailman, J. P., and Strier, K. B. (1997). *Planning, proposing, and presenting science effectively.* Cambridge, U.K.: Cambridge University Press.

Oxnard, C. E. (1983). *The order of man.* New Haven and Hong Kong: Yale University Press and Hong Kong University Press.

(Smith and Jones (1893) is fictitious, as are rabbitoes, in case you had not guessed.)

6

Advanced Technology
Observation Using a Computer Data Logger

Over the period from 1970 to the present, and despite the persistent use of pencil/pen and paper notebook, there has been a progressive increase of interest in and utilization of data logging devices. These devices perform two important functions:

1. They generally restrict the format and type of data collected by the observer, intensely focusing attention on clear categorization of behaviors, and
2. As Forney, Leete, and Lindburg (1991) note, they provide shortcuts between the collection and analysis phases of research.

With data collection proceeding with paper and pencil, or especially with a tape recorder, many people display a tendency to be overly descriptive in their observations, thus building up a large volume of "unique" details that cannot be readily analyzed by statistical procedures. There also arises a significant problem in translating, condensing, and eventually coding the mass of data for computer-based analysis (it is assumed that analytical methods *will* involve statistical treatment and the use of computers in order to shorten the process). Solutions to these problems have been, in some form, available since the late 1960s. Field equipment capable of collecting data in numerical coded format and transcribing the dataset to a mainframe computer have been available since the 1970s (Torgerson, 1977). Most of these systems have been expensive, fragile, and complicated to operate. Interest in and examination of the SSR system (Stephenson, Smith, and Roberts, 1975), and later experimentation with the KIBOS program (Lifshitz, O'Keeffe, Lee, and Avery, 1985) served as stimuli in the development of observational data logger systems and for the programs described below.

The advent of the laptop and palmtop notebook computers made the recording of data under field conditions much easier for both anthropologists and primatologists (Dyson-Hudson and Dyson-Hudson, 1986;

Wood, 1987). A number of purely software approaches have been developed as flexible data collectors for primate studies (A. Coelho, personal communication, May 1988; K. Glander, personal communication, April 1991). However, there have been commercial devices whose sole reason for existence was collection of data, but relatively few of these have been of significant interest to field primatologists. Among these products were the products of the Electro-General Corporation, which produced the Datamyte series of data loggers (Torgerson, 1977). The Husky Hunter (an MS-DOS/Windows compatible), a more general type of computer designed for harsh environments and rough use, has been available for many years through Forestry Suppliers Inc. This unit looks much like a Tandy Model 100/102 and uses a similar screen, but it is waterproof and environmentally hardened. It is also rather expensive.

With many of these devices, working on arboreal primate species in a rain forest would require a hand to hold the device, one to type/enter data with, and one or two more to manipulate binoculars. Since it is unlikely that multihanded observers exist, these devices are not very practical in many field situations. In less physically stressful observing conditions, such as captive circumstances with a fixed observation blind, laptop and palmtop computers can be effectively used. The most practical, useful, and inexpensive field computers over the 1980s were the Tandy/Radio Shack Model 100 and Model 102, although for most outdoor situations they needed a Safe-Skin or similar plastic cover to improve their environmental hardening. At Calgary, our research and instructional programs still use these computers, and they can still be found on the used market in good condition at quite low prices.

Currently, there are a number of other general purpose but relatively small devices, specifically the palmtop computers including the Psion Organizer series or its much tougher cousin the Workabout, and even the Palm series of pocket organizers. But a number of the small computers from Hewlett-Packard can serve effectively as data collectors. The major problem with most of these devices is structural fragility and a lack of environmental hardening, coupled with limited ergonomic efficiency in field use. All this being said, researchers have employed Hewlett-Packard HP95LX computers placed inside Ziploc plastic bags to collect data in tropical rain forests.

In considering the use of a computer for observational recording, the following five points should be considered (expanded from Lehner, 1996).

1. Is the unit able to do the job? Does it have sufficient memory to hold the expected amount of data that will be collected? Is it programmable? Are there available programs suited for the collection routine and perhaps performance of the required analyses?
2. Is the unit appropriately portable? Will the process of data entry interfere excessively with the maintenance of observation? Will the weight of the unit cause excessive fatigue in the observer?
3. Is the unit sufficiently hardened against environmental damage? It should be waterproof or, at least, water-resistant, and able to operate

beyond the temperature and humidity extremes of the study location. It must also be resistant to physical damage.

4. What do you do for repairs? Is it readily serviceable or replaceable? Is service/replacement available near the research location? Is the manufacturer able and/or willing to provide quick mail or courier repair service at reasonable cost? Is the price such that a spare can be taken to the field?
5. What do you do if your machine becomes inoperative and nonreplaceable? Is there a backup research design and/or data-collecting procedure?

In the end, the selection and implementation of a computerized data gathering system is very much a product of personal evaluation and preference. It is wise to remember the twin principles of " No research plan survives contact with the study species" and Murphy's Law, "If anything can go wrong, it will." Plan for problems, expect them, and be prepared to revise plans as necessary.

As noted above, common and inexpensive notebook systems from the 1980s were the Tandy Model 100 and Model 102 computers. This basic computer has also been the machine of choice for development of bar code–based scoring systems (Forney, Leete, and Lindburg, 1991). While a bar code–based system has some advantages in terms of efficiency and accuracy, it does have an incremental startup cost, is more consumptive of battery power, and may prove less rugged in use under field conditions. Under normal operational conditions, the Tandy 100/102 computers use up a set of alkaline batteries in about twenty hours. Substituting rechargeable nickle-cadmium batteries is less useful as they provide only about two hours usage on a charge. Modern palmtops, like the Psion Workabout or the HP360 series, run for months on their batteries.

The software system described here was intended to perform effectively under most field conditions for the two observing protocols, focal animal and focal time sampling. The CD contains working copies of both DATAC-6 and the DATAC-7 program for several different computers. The most recent version of the program is available from the programmer's Web site:

http://www.PrimateData.com

DATAC 6

DATa ACquistion Six was the sixth incarnation of a series of hardware and software data collectors that have been under development at Calgary since 1978 (Paterson, 1988; Paterson, Kubicek, and Tillekeratne, 1994). This incarnation was purely a software version for the Tandy 100 and 102 computers running under the BASIC interpreter. A complete listing of the program in ASCII text can be found on the CD, or it can be downloaded from the PrimateData Web site. This listing must be loaded into the Tandy and BASIC used to convert it into an operational program.

DATAC 7

DATa ACquistion Seven is the seventh incarnation of this system and is intended as a flexible system for a large range of platforms. The program has been completely rewritten by Donald Vandenbeld to be cross-platform-compatible. The most recent version runs under the Java environment.

The purpose of this program is the collection of either continuous or interval sampled data, and the organization of that data into an analyzable file. Timing control is part of the program, as is calculation of durations during focal animal sampling. Program operation is straightforward and can be achieved by following the instructions and prompts as the program operates.

Focal Animal Sampling Subprogram

The focal animal sampling subprogram is intended to allow the collection of free form, continuously sampled observations. The program prompts for data items, which are stored as the file header along with specific details for each individual sample. A series of samples, normally a working day's data collection, is stored in a database. The sample header data is used as the tag for locating samples within the database. The program begins to time the sample from the first keystroke of observational data made by the operator. Observations can be recorded in any form. Input of data may be continuous up to 255 characters (approximately six and one-half lines of data). Different variables are separated by tabs. The line of data ends with a return, upon which the program waits for the next data line. It is important to recognize that the timing algorithm waits until the first keystroke of the new line before calculating the duration of the behavior in the previous line. This means it times from the first keystroke of line one until the first keystroke of line two, calculates the elapsed time in seconds, and places that value into the previous data line as the duration of the behavior. It continues to perform this action until the sample period runs out, whereupon it performs the last calculation and ends the sample.

Focal Time Sampling (FTS) Subprogram

The FTS subprogram was originally intended to operate under the focal time sampling rules of Baulu and Redmond (1978) described in exercise 8, but it can be used for the collection of any form of interval data. The program controls the timing of the activity, providing both a visual and audible prompt, but the latter can be switched off. The program will prompt for the information to create a file header in the same fashion as the focal animal sampling subprogram, and proceed to prompt for the sample header. It will beep or not beep, according to the established setting, and prompt for the input of a set of variables at each interval. As with the other subprogram, up to 255 characters of data may be entered before hitting the return, and it tabs between subject and variable pairs. If no input is made, the program automatically proceeds to the next interval, leaving the input blank. At the completion of the sample period, the program will close the file.

As noted, the program can be used for standard scan sampling (Altmann, 1974). However, there is an obvious limitation on the size of the group in conjunction with the interval chosen, since the input must be completed before the next interval beep. The operating procedure would be as for a normal FTS file, but each subject and activity code would be entered in sequence at the prompt, and the series ends with a return.

Some General Considerations

The files produced by DATAC 7 form a database structure from which the included data organizer component can produce reports independently or transfer information to a word processor, spreadsheet, statistical or communications program. DATAC 6 files are simple ASCII text files and can be employed by virtually any program. The Tandy version (DATAC 6) of the program occupies substantial memory when operating. In a 32K RAM Model 100, the program will leave sufficient space for 12 to 13 kilobytes of files if the floppy disk driver is loaded, or 16 to 17 K if storage is on cassette tapes. Naturally DATAC 7 can handle much larger files, up to the capacity of the hard drive, zip drive, or other storage device.

Exercises

Exercises appropriate to this technology level would be a repeat of the focal time sampling or focal animal sampling exercises. The utilization of a computerized data collector in no way changes the characteristics and protocols of either form of study. Conduct of these exercises is considered to be advanced-level work since it is reliant upon access to computer equipment. It is suggested that while data collection can be carried out effectively with a data logger, the analysis phase is most efficiently performed on a laptop or desktop computer using appropriate statistical software.

7
Field Gear, Equipment, and Accessories

Any fieldwork project requires careful budgeting for equipment, observational tools, clothing, antimalarial and other drugs, and a multitude of gear needed to survive and work in a bush or forest environment. While this workbook is not intended to cover the necessities for fieldwork in detail, it is considered appropriate to make some recommendations that may help to avoid problems.

First, some books that will be of value to planning a field project.

Werner, D., Thuman, C., and Maxwell, J. (1992). *Where there is no doctor: A village health care handbook* (2d ed.). The Hesperian Foundation, P.O. Box 1692, Palo Alto, CA 94302, U.S.A.

Wiseman, J. (1986/1996). *The SAS survival handbook: How to survive in the wild, in any climate, on land or at sea.* London: HarperCollins.

Barnett, A. (1995). *Expedition field techniques: Primates.* Expedition Advisory Centre, Royal Geographical Society, 1 Kensington Gore, London SW7 2AR, U.K.

National Research Council. (1981). *Techniques for the study of primate population ecology.* National Academy Press, 2101 Constitution Ave. NW, Washington, DC 20418, U.S.A. (out of print)

Lehner, P. (1996). *Handbook of ethological methods* (2d ed.). Cambridge: Cambridge University Press.

Krebs, C. J. (1999). *Ecological methodology* (2d ed.). Menlo Park, CA: Addison-Wesley Educational Publishers.

The first volumes in this listing provide two of the most concise and practical collections of medical and survival information currently available. While much of what they contain is irrelevant to primatology studies, the material can be of great value and serves as a solid practical introduction to some of the problems that can be encountered in field conditions. Many will reject Wiseman merely because of its military ancestry, but this is unwise as much of the field experience of soldiery is directly applicable to the circumstances of a primate observer. Look for the practicalities and keep an open mind.

The latter four volumes supplement and complement this workbook. They provide in-depth coverage of many aspects that are not treated here, as well as providing methodological tools to expand the research plan.

The National Research Council book provides a brief and succinct "Checklist of Essential Equipment" which is modified and restated here.

Basic Field Work
- ❑ Binoculars
- ❑ Compass
- ❑ Stopwatch
- ❑ Watch
- ❑ Notebook
- ❑ Pen (indelible ink)
- ❑ Knife

Mapping
- ❑ Tape measure
- ❑ Compass
- ❑ Surveyor's tape
- ❑ Rangefinder

Animal Marking
- ❑ Traps and bait
- ❑ Dartgun and darts
- ❑ Blowgun and darts
- ❑ Crossbow
- ❑ Collars
- ❑ Tattoo Equipment
- ❑ Freeze branding supplies
- ❑ Anesthetic
- ❑ Scales
- ❑ Calipers
- ❑ Measuring Tape
- ❑ Scissors
- ❑ Dye

Field Data Analysis
- ❑ Graph paper
- ❑ Ruler
- ❑ Calculator

Plant Identification
- ❑ Plant press
- ❑ Corrugated aluminum
- ❑ Cardboard sheets
- ❑ Newspapers
- ❑ Tree tags

Naturally, the list of essentials depends very much upon the research design, the species involved, and the political issues in the country where the work is to be carried out. Many countries will require special permits for the importation and use of dartguns, blowpipes, and crossbows. And today much of the field data analysis would be conducted on a laptop or palmtop computer, which may also replace the stopwatch and notebook under some circumstances.

TRANSPORT

In any recent field research project, the major expense is transport to the field. It is important to recognize that travel is much less than the bargain it once was (long gone are those airfares to Europe at under a hundred dollars!), and consequently, a consideration in any project is simply the cost of getting there. Nor can one expect that transport within the host country will be easy. There are still areas in the primate study world where a "full-pack" hike of more than a day is necessary to get to an established research site. If the project requires more gear than can be comfortably carried on your own back, there will be additional costs to hire porters, who will also have to be fed and sheltered. Most countries have some form of local bus service that varies in quality from good to horrid. A field researcher needs to be flexible in his or her transportation planning and prepared to tolerate a lot of discomfort in order to get to a field location.

DISEASE PROPHYLAXIS

Fieldwork in tropical conditions involves exposure to a broad range of parasites, bacteria, and viruses that will be new to a temperate-zone resident. Many of these will be rather benign or produce only short-term illnesses, but some—malaria, schistosomiasis, onchocerciasis, and a range of developing new viruses (Lassa, Ebola, Kayasonur, Marburg)—are more serious, potentially lethal.

In general, the most appropriate prophylaxis is avoiding exposure to the pathogen in the first place, but for the most common—malaria—chemical defenses are available. While the medical profession in North America often seems to be united behind the use of mefloquine (sold under the trade name *Lariam*), there remains a great controversy about its effectiveness and side effects. There have been frequent reports of psychological disturbance, hair loss, restless dreaming, and a number of other complications. Some fieldworkers, myself among them, prefer NOT to use Lariam, opting instead for Doxycycline—an antibiotic-derived compound that is, perhaps, less effective but is much better tolerated. Of course, individual responses vary, different medical associations make different recommendations, and many workers on long-duration projects elect to avoid prophylactic drugs altogether and instead use Halofan or Artemisia as a treatment when malaria does appear. Wearing of long-sleeved shirts and full-length trousers, as well as sleeping under a pyrethrum-treated mosquito net, are additional safeguards against malaria.

Avoidance of exposure to *Schistosoma mansoni* and its related species is the most practical prophylaxis as the treatments for the resulting disease (bilharzia) are both dangerous and debilitating. Since the snails that host the infective larval stage live in standing and slow-moving water, exposure can be avoided. In part, schistosomes are the reason all drinking water should be run through a ceramic filter system at minimum, and preferably boiled before filtering.

Onchocerciasis, also known as river blindness, is associated with *Similium* flies that live near flowing water and transmit the microfilarid worms when they bite. Many areas of Africa, Asia, and Central America host the disease, but fortunately the infective rate is low. Treatment is a single dose of the drug Ivermectin taken when leaving the infested area. This kills the juvenile worms, and there should be no further problem. If the worms are allowed to mature, they become far more difficult to deal with and can only be suppressed, not eliminated. In many temperate-zone countries Ivermectin is not authorized for human use, but this is primarily because there are no *resident* parasitic diseases of humans that require it. However, due to its importance in tropical environments, this policy should not be adhered to rigidly.

Clothing

What to wear for fieldwork? While this is a rather personal decision, there are some important considerations. In many countries it is NOT appropriate to wear camouflage-pattern clothing under any circumstances, as this can lead to a researcher being mistaken for an insurgent or guerrilla. It is safest to avoid military-style clothing even though it may seem to be perfectly suited to the conditions. Some research sites even specify the type and color of clothing that workers are required to wear, but a good practice is to keep to browns, blues, or greens and avoid loud colors or prints—save the latter for parties and meetings. My own preference in the field runs to blue jeans and denim long-sleeved shirts, and a Tilley hat. (Tilley Endurables Limited, a Canadian company, produces a wide range of travel clothing that purports to be mildew-proof and very tough—their washing instructions are "give 'em hell!")

Footgear

There is one priority rule for any fieldworker: *Take care of your feet!* They are your most important asset and the means by which all work and travel become possible. I have personally broken toes and wrenched ankles on occasions in the field. I have known students to severely sprain ankles and even contract trench foot under field conditions. Good-quality, sturdy, well-broken-in hiking boots may be the best field equipment that you can spend money on, but there are conditions under which they will deteriorate very rapidly in the tropics. One colleague reports that his lasted for only six months. In my own case, as I always seem to work in forests, my normal footgear consists of a pair of good-quality gumboots of North American manufacture. The Bata Shoe Company locally produces gumboots in many

African countries, and while adequate, they are not as sturdy as the European and American product. Part of the reason for wearing gumboots is the presence of safari ants, driver ants, or soldier ants, which have the distressing pattern of climbing a fieldworker's body and biting ferociously. Gumboots are slick enough—as long as they are kept clean—that the ants cannot obtain sufficient traction to climb. If hiking boots are worn, then pant legs must be tucked into socks to keep the critters from climbing up inside the clothing. An additional tip from an old oilfield worker with experience in Sumatra was to use duct tape to seal the pant legs to the boots—especially useful where leeches are common.

PACKS

Fieldwork seems invariably to involve the transportation of a multitude of devices and materials needed during the working day, including the all-important lunch and water bottle. There are two considerations for the method of carriage: (1) what is the weight of the equipment? and (2) is a particular piece of equipment required to be available immediately? The majority of fieldworkers favor a small daypack to transport the day's needs conveniently and comfortably. However, equipment tends to get buried in the pack and sometimes several minutes are required to locate, exhume, use, and then restore it. An option is to have a number of small pouches or containers fastened to a heavy belt, each device having a separate and regular location. This tends to speed up access, but many people find the belt system cumbersome and irritating to wear for long periods. Perhaps an appropriate solution is to employ the U.S. Army field gear known as "Alice." For those who have never seen it, this system involves a web belt with an attached set of padded suspenders, a multitude of various-sized pouches that can clip to the belt, and a daypack that integrates with the suspenders. And there is a quick-release buckle available for it.

MACHETES/PARANGS/BUSH KNIVES

Some fieldworkers prefer not to carry a bush knife of any kind, and that is their personal decision. On the other hand, many, like myself, prefer to do so. The choices of knives are varied and may range from the traditional Collins machete issued to the U.S. Marines to a handmade Malayan parang (as seen in the hands of John Ellefson [April 1979]), but tend to fall into two categories. The first is the yard-long machete or panga, which is available throughout the tropical world, though it may go by a multitude of local names. It combines a lengthy reach with sufficient mass to readily chop through thick vegetation, but, when not in use, it represents a significant load on the hip or back. Like many other fieldworkers, I prefer a shorter knife, about 30 centimeters long. My own is of the Indian Army/Gurka regiment kukri pattern, which is light enough to avoid being an impediment and still of sufficient size and weight to cut brush when needed, and could even be pressed into defensive use.

A bush knife can also be used as a hammer replacement to fasten tags on trees and drive stakes for survey plots. Take note, however, that some countries take a dim view of bringing such devices within their borders and may confiscate them at Customs even though they may be readily purchased inside the country.

Binoculars

A good pair of binoculars is almost an identifying trademark of the experienced field primatologist. Binoculars are necessary observational tools in almost all aspects of primate fieldwork, and since so much time is spent with these devices in front of one's eyes, it is appropriate and wise to obtain the best! Top-line instruments come from Leitz (Leica), Zeiss, Swarovski, and Nikon. The first three only manufacture top-line equipment, while Nikon produces several distinct grades of quality. Basically, a top-quality pair of binoculars for fieldwork should be extremely rugged in construction, have superb-quality glass lenses with top-grade multicoatings to reduce reflections and ghost images, normally come equipped with fold-down eyecups that allow for use with and without prescription glasses, and may be dry-gas filled as well as hermetically sealed. Such products are indeed expensive, typically over $1000 US, but also come with a lifetime warranty. While many beginning observers elect to use bargain basement or department store binoculars, the experiences and clarity of vision are disappointing. Eventually everyone wants to move up to quality equipment.

The issue of objective size and magnification always arises in discussions of binoculars. The two numbers associated with binocular descriptions represent the magnification as 6, 7, 8, or 10 "x" (times magnification), and the diameter of the objective lenses in millimeters. Thus, a common size for observational use is 7 x 40, 7 x 42, 10 x 40, or 10 x 42. Initially, a student might expect that the most commonly available size (7 x 50) would be better, however, while the 50 millimeter lenses do gather more light, the penalty in extra weight is generally not worth it. Other issues that ought to be considered in the selection of binoculars are the matters of the field of view—usually the wider the better—and the diameter of the exit pupil. The exit pupil is the light disk that is the focus of the eyepiece, and generally it should be the same size or slightly larger than the diameter of your own pupil. This, effectively, means that it will be easy to find the image, and not too much will be wasted. All of the best-quality binoculars will have these features optimized for the normal range of human eyes. Binocular manufacturers maintain their own Web sites providing details and data on their products; however, the Swarovski Web site has a superb introduction to the technical features, advantages, and limitations of the various binocular designs.

Cameras, Lenses, and Film

Photography of wildlife, both faunal and floral, are specialized topics within the discipline and frequently require specialized techniques. The details of these can be found in innumerable photographic manuals and will not be treated here. Tropical fieldwork conditions place substantial strains upon any camera equipment, and it seems to be especially hard on the modern electronic wonder cameras—in my experience they die with great regularity. It is for this reason that I am somewhat anachronistic and continue to use Nikon F and F2 cameras manufactured in the 1960s and 1970s. In any case, the primary need in field photography is an interchangeable-lensed "single lens reflex" camera and at least two lenses. The first and most useful is the normal lens—of 50- or 55-millimeter focal length—that most frequently is a part of the standard commercial package. This basic lens can perform at least 50 percent of all needed photography, and if it has a Macro, or extreme close-up capability, can handle all of the basic biology recording. The second lens should be a long telephoto or telephoto zoom with at least 300-millimeter focal length at the upper limit. This is the workhorse for photos of the primates themselves—up in trees, across the river, and even at distances of 5 to 10 meters—as it enables you to obtain a larger image on the film. The penalty for any long lens is the weight and size. There are a number of new optical designs that are much shorter and lighter in weight and should be investigated before making a decision.

What choice of film? While print film is the most popular form in recent years, slide film is still the most important medium for field researchers. The reason for this is that the pictures will most likely be used for lectures and presentations to groups rather than being in an album to be passed around. It is probably most appropriate to employ both types, with the understanding that the end results will serve very different functions. Of greater importance may be the fact that slide or transparency film is not always available in some areas of the world. Of even more significance is the speed rating of the film, usually given as an ASA or DIN value expressing its sensitivity to light. For the purposes of someone working in an open savanna environment, it may be perfectly appropriate to restrict film selection to a single rating such as 200 ASA. But if the photography must be done under a forest canopy, high-speed film is a necessity. Fortunately, there are now broad ranges of both print and slide film in the 400 to 3200 ASA range. A problem is the need for storage at cool temperatures that many of these types of film require—not always feasible in the field. The quality degradation can be extreme at high temperature. Hence, the recommendation is to take a variety of film types and try to have the fastest film that does not require refrigeration as your main supply.

An obvious alternative would be a digital camera system, and while many of these are now approaching the quality of a good 35mm system, they have not yet reached that level. They also come with all of the attendant "electronic" problems in the tropics and need batteries, chargers, a supply of disk or flashcard media, and so forth. Yet these problems will undoubtedly be overcome within the next several years, and fieldworkers will become more electronically oriented in all of their activities.

Video

Modern video cameras, particularly the Hi-8 equipment, are capable of producing broadcast-quality imagery, but as with any other highly technical craft, a great deal of experience and practice is required to take advantage of these capacities. Video can be an extremely attractive adjunct to a field study, and has the potential, if quality imagery is collected, of recovering some of the field expenses. A major caveat is that most people, even the majority of field researchers, are not cinematographers, and the imagery is not likely to be saleable. If the objective is to use a smaller handycam to document interesting behaviors, and the weight of camera, tapes, and batteries can be accommodated, then video may serve a purpose. However, for most of us, we are better off leaving the production of a cinematic epistle to the professional videographer. For a visual documentation of the costs, benefits, and problems of such work, a viewing of Alan Root's film *Lights, Action, Africa!* is very worthwhile.

Tape Recorders, Microphones

The primary purpose of tape recorder usage in the field should NOT be for taking observational notes. Using a tape recorder in this fashion is an invitation to the introduction of significant data-reduction problems. It is difficult to enforce protocol and record only what is needed for the behavioral record. The real utility of the recorder is for the collection of primate vocalizations in sufficient quality and quantity so that they may be subjected to a laboratory-based sound analysis system. There are many directions in which such work can go, and these are topics beyond the scope of this workbook. The acquisition of good recordings depends upon a number of acoustic conditions as well as the technology employed. While I cannot advise on the acoustic situation, the technology at this time is quite straightforward. I have been advised that the best portable recorders for field use are the Sony Walkman Pro (Model WM-D 6C) and the Marantz PMD (200 series). Both are cassette recorders of reasonable size and weight and have the capacity to record over the entire human-audible range, when good quality metal tape is used. The recorder is not the most limiting machinery, the microphone is. Most microphones are remarkably limited in their sound-capturing capability, and for field recording of vocalizations a "shotgun" type microphone is virtually mandatory. In the range of narrow-angle "shotgun" devices available, the most highly recommended is the Senheiser MKH 416, at approximately $1000 US, or one of their K-6 system versions with an ME-66 shotgun capsule at about $500 US. Either model will enable the recording of vocalizations emanating from the canopy with superb clarity.

GPS—Global Positioning System

Global Positioning System receivers are getting better each month, and once the degradation of the signal is eliminated, the accuracy will be within a meter in nearly all areas of the world. But until some of the problems of dealing with obtaining a clear set of satellite signals under a canopy are solved, these units are marginal for forest workers. Some work better than others, so a degree of effort is required to find a unit that will suit the needs of a fieldworker. The primary relevance of GPS for field research is mapping, although a few studies have made use of the data for group dispersal and dynamic studies. Naturally, there are some working situations that will mandate the use of a GPS unit. The major problem, as alluded to above, is the limited sensitivity of current receivers when under a forest canopy. Personal experience with this situation leads to two solutions: finding a relatively large opening in the canopy and remaining there long enough to acquire the necessary three satellites, or placing the receiver as high as possible and leaving it for fifteen to thirty minutes to maximize the opportunity to acquire signals.

Data Loggers

As detailed in the previous chapter, the utilization of a computerized data collection system is now feasible in practically all environments. The key features of such systems are: a hermetically sealed or at least waterproof unit; very low power consumption (so that battery supply does not become an issue); and simplicity of data storage and transfer to central computers. Based on these criteria, the most practical, but relatively expensive, device is the Psion Workabout. This unit is now an integral part of The Observer package distributed by Noldus, although only a well-supported individual can afford this system. Alternatives seen in the past have been the computers built for forestry use, the military, and the very few developed for biologists. The cheapest expedient may be the one that a Brazilian fieldworker showed me: A Ziploc plastic bag containing a Hewlett Packard HP95LX palmtop. As noted in the previous chapter, it is always necessary to allow for the failure of the technology and have backup plans ready for implementation.

Solar Chargers

Solar charging systems have been the savior of many field stations and research projects, but it is necessary to be aware of their limitations. The two available types of photo-voltaic cells are quite different in structure. Most common is the solid crystalline form usually bonded to a glass plate, while the newer "amorphous" crystal type is bonded to a thick plastic substrate and can be rolled into a tube for transport. The glass plate forms currently cost $70 US per square foot, while the amorphous crystal forms are somewhat more expensive. Both varieties require some caution in wiring as they can be easily damaged. Commercially manufactured systems are available from a number of sources in Europe and the Americas that sell primarily through their Web sites. Like a battery, photocells can be wired in series to provide higher voltage, or in parallel to provide higher amperage, but the number of units needed to power a major appliance or recharge a large battery set would be difficult and costly to transport into the bush. It also should be recognized that the power available depends upon the intensity of sunlight. Cloudy days or shaded installations mean less power, less recharging of batteries, and more circumspect power usage during the evening.

Microscopes/Magnifiers

Permanent research sites might have a research microsope present, but most probably do not, and the feasibility of taking one into the field is limited. Most field needs can be accommodated through the use of a folding hand lens of 10, 15, or 20 X magnification and can be included in a small pouch as part of the field gear. If higher magnification is required, there are a number of field microscopes available at prices from $60 US for a Panasonic to $700 US for the Swift Field Microscope (an innovative and effective tool in a remarkably small package). These would generally be necessary only when examination of smears or slide preparations are important to the research design.

Ecological Equipment

A multitude of useful and specialized equipment could be listed under the heading of ecological equipment, and most of the main categories can be found in the table of recommendations from the NRC checklist earlier in this chapter. A good-quality compass is a necessity for every fieldworker, and there are many available, but the Brunton Pocket Transit is a device that builds upon the basic compass to include an inclinometer and a plane-table transit. It is expensive but serves as three distinct instruments in a very compact unit. Rangefinding and inclination are two capacities that have been served by tape measures and inclinometers for the past century, with the

Relascop and Sunto Clinometers being the standard forestry instruments for measuring the height of trees. To a great extent, these instruments can be replaced in a field kit (for a fraction of the cost) by a laser rangefinder. One simple, but expensive device, the Crown Densiometer, measures the percentage of canopy or crown vegetation over a fixed location. This is a spherical mirror with a box grid engraved on its surface, and when used as directed will produce a consistent measure of the overhead obstruction.

In forests it is often necessary to tag trees for any one of a number of research purposes, and the forest industry has developed a large variety of aluminum tags and even aluminum nails for these needs. Ethically, the use of aluminum materials avoids damage to the tree, and if the wood is ever delivered into the hands of sawyers, there is no danger of shattering a powered sawblade—an occurrence that can be fatal to the operators if a steel nail is encountered.

Repair Materials

Many things break or wear out during the course of a field study, and while some cannot be repaired by the researcher, many others can be. Two important materials should be on hand. A standard for all handy-persons, duct tape can hold many things together, provide sealing for packages, act as emergency bandages, and even serve as tree tags when needed. Purchase the best brand available, because much that is marketed today is not as good as the original. A high-quality option is gaffer's tape, the material used in Hollywood and ubiquitous on movie sets. The second material is a tube or two of Goop. This is a general purpose adhesive that is available in varieties tailored for use on footwear and as a general household adhesive. This material is ideal for sealing tears in gumboots, patching rain ponchos, sticking up signs at trail intersections, anywhere that adhesive or patching is needed.

Snakes and Ticks, Scorpions, and Spiders

Beginning fieldworkers are naturally very concerned about the dangerous lifeforms at their field site, and indeed many snakes, a few scorpions, and some spiders are dangerous enough to warrant serious consideration. Yet, the fact is that poisonous reptiles and insects are not out there to make a meal of unsuspecting researchers. Given a choice, they will flee more often than not. The danger level then descends to the few that demonstrate overtly aggressive manners (mambas in Africa, bushmasters in the Americas) or are so well camouflaged that it becomes possible to inadvertently step on them, thus provoking a retaliatory strike. Some people seem to attract snakes, encountering them frequently; others, like myself, seem to encounter them rarely—yet I once came within centimeters of stepping on a Gaboon viper (a species for which there is no antivenin).

To protect oneself, it may seem appropriate to have snake leggings such as the many varieties available from Forestry Suppliers, Inc., but many of the field assistants will give you that look that says "yet another crazy North American." For myself, I wore leather snake leggings during a single study in southern Mexico where there were large numbers of Fer de Lance snakes, but otherwise have left them at home. Every research site should, in spite of this perception, have available a snake bite kit and appropriate antivenins. The latter are often supplied in two vials and the task of mixing them together and managing the injection into a vein is not for the faint of heart, nor is it something that one wants to do in the field during those shaky moments after having been struck. Antivenins are best left in the base camp as they often need refrigeration. What is more suitable in an emergency kit are the Insect/Snake Venom Extractor devices and Tick Removal equipment manufactured and sold by Sawyer Manufacturing Company. They can be reached at venom@icomm.ca or (718) 227-6234 as their products are often difficult to find in local suppliers. (Thanks to Steve Grenard of the Herpmed list for this information.) The best prophylaxis to avoid dangerous situations is to be vigilant and aware of your surroundings at all times. This rule applies for all venomous organisms. However, many snakes and spiders are also brightly colored and quite beautiful—I particularly remember a rhinoceros horned viper in the Budongo Forest who was resting at shoulder height on a tree branching. Beautiful but deadly!

Wills

One final matter drives the field researcher into the legal offices. Everyone undertaking field research, whether it is a brief survey or a year in the bush, needs to have a completed legal will on file. This does nothing for the research but certainly can make life simpler for family and heirs or offspring. A visit to the lawyer and updating of our wills has always been one of our important steps before a trip to the field.

Final Thoughts

The conduct of field research in a distant and undisturbed habit may initially sound exotic and inviting, but the realities of the experience can be very different. If one goes off with the expectations that everything will be perfect, that the local humans will be accommodating and respectful of your ideas and intentions, that you will have all the appropriate gear, and that everything will proceed smoothly from start to finish, then the shock of reality may be unnerving. For the truth is, research is not like a fairy tale; things do not always go right and the axiom "No research design ever survives contact with the subject species" is true more often than not. I have had the experience of living through a coup d'état; I have broken bones, sprained ankles, been bitten, stung, and damaged by both plant and animal life; I have seen my research design crash with both device failure and subject animals not behaving anything like they were supposed to. In short, expect culture shock, and problems with people, equipment, materials, and your subjects. Yet, in spite of the travails of failed grant applications, failed research designs, and all the other irritations, I would not have chosen a different course for my life. If you choose to take a similar path, I hope it proves enjoyable and rewarding.

Part 2

Exercises

The Log Book

A log book is a normal feature of any research enterprise. It constitutes a formal and legal record of the research work that has been undertaken and serves to record experimental designs, research objectives, new insights, materials used, and so forth. Indeed, the log book constitutes a journal or diary of the activities of a laboratory or an individual researcher. To confirm that log books in some form or another have been important in the history of science, I need only point to the famous notebooks of Charles Darwin, Leonardo da Vinci, Ramanujan, and Isaac Newton. All have been published and reflect the cardinal importance of these scientists in the development of Western thought. Laboratory log books in the physical sciences have, in recent years, even served as legal evidence in cases relating to priority in establishment of patent rights and in academic dismissal proceedings after accusations of plagiarism and data falsification.

A log book serves to maintain a record more precisely than human memory is able to do, and thus serves as a valuable resource in tracking experimental manipulation and the development of theory. While the log book is not usually a published document, it will often serve to hold the outlines and concepts as well as data that later leads to publication. Maintenance of a log book is a habit that all researchers are advised to establish early in their careers. As students proceed through assigned exercises, it is recommended that they keep a log book. Periodic review of its contents will remind the reader of past activities or observations relevant to some current problems or will serve to correct drift in their questions or hypotheses. Some instructors may require the keeping of a log book for grading purposes.

A log book is a personal document or, at an organized laboratory, an official one, and no two are alike. Consequently, my views will differ in many ways from those of other researchers. As a field researcher, I have always maintained two log books: one for accounting purposes in which every expenditure is noted as to date, what it was for, and the amount involved—

usually in local currencies—and a second for research-related material. Data that may be relevant include the date (normally set off in some way to serve as an index marker), times of entering and leaving the working area, times of contacting and leaving or losing the subjects, weather conditions (temperature, rainfall, relative humidity, winds, etc., according to what instrumentation is available), summaries of interesting occurrences, general troop movement, activities, food materials, and perhaps preliminary data analyses. When the research work involves activities such as tree sample plot surveys, phenological examinations, or reading of local materials not available elsewhere, the data so collected tends to end up in the log book. It has become my practice to keep the log book on my person when traveling; the risk of loss is lower than if it were to travel in checked baggage.

Almost any form of notebook can be used as a logbook, however, a bound volume—the laboratory notebooks used in science courses are excellent—has the advantage of being flatter and less likely to lose pages. I have used everything from 3″x5″ miniature notebooks to 8″x10″ lab notebooks. My wife has consistently preferred commercial "page-a-day" type journals.

Other researchers will have different views on what a log book is, what it should contain, and even the importance of such a document to their personal research, but this affirms the variety of perspectives in the research endeavor.

Exercise Control

The next pages provide "control" sheets to specify which exercises are required in your course and the components that need to be included in each. The second control sheet is a check sheet to verify that all of the required components are present in a project before it is submitted for evaluation. The third is intended for use with the ecological exercises in a field school. Separate files for each sheet are present on the CD.

Exercise Control Sheet

(Duplicate as many as needed)

Preliminary Studies

- ❑ Identification
- ❑ Ethogram
- ❑ Observation Schedules

Group Comparative Studies

- ❑ Behavior Profile
- ❑ Time Budget

Major Methodology Practice

Method	Report Form	Analysis	Statistics	Illustration
❑ Scan	❑ Full Scientific Report	❑ By Behavior	❑ Mean	❑ Graph
❑ One/Zero	❑ Brief Report (2- to 3-page)	❑ By Age-Sex Class	❑ Standard Deviation	❑ Interaction Matrix
❑ Focal Time	❑ Labeled List	❑ By Age or Sex Class	❑ Median	❑ Sociogram
❑ Focal Animal		❑ By Individual	❑ Mode	❑ Table
❑ All occurrences			❑ Frequency	❑ Image
❑ Matched Control Pair			❑ Relative Frequency	
			❑ Rate	
			❑ Hourly Rate	
			❑ MDB	
			❑ MDH	
			❑ MDI	
			❑ % Time	

Special Studies

- ❑ Inter-Observer Reliability
- ❑ Postural Congruence
- ❑ Spatial Locational Study
- ❑ Nearest Neighbor Study

Report Control Sheet

(Use to verify that all required components are present)

❑ ***Brief Report*** ❑ ***Labeled List*** ❑ ***Full Scientific Report***

❑ ***Title Page***

- ❑ Name and ID
- ❑ Course and Instructor
- ❑ Title of Project

❑ ***Introduction***

❑ ***Subjects and Methods***

❑ ***Results***

- ❑ Graphs
- ❑ Tables
- ❑ Interaction Matrix
- ❑ Sociograms
- ❑ Illustrations
- ❑ Statistics

❑ ***Discussion***

❑ ***Conclusion***

❑ ***Was the Research Question answered?***

Field Ecology Exercises

- ❑ ***Population Surveys***
  - ❑ Line Transect
  - ❑ Quadrat Survey
- ❑ ***Home Range Survey***
  - ❑ Quadrat Use
  - ❑ Least Polygon
  - ❑ Home Range Program
- ❑ ***Sample Vegetation Plots***
  - ❑ Plotting
  - ❑ Height Determinations
  - ❑ DBH
  - ❑ Ecological Indices
- ❑ ***Phenological Survey***
  - ❑ Transect Method
  - ❑ Quarter Tree Leaf, Flower, and Fruit
- ❑ ***Vegetative Sampling***
  - ❑ Voucher Specimen
- ❑ ***Faecal Analysis***
  - ❑ Wash and Examine
    - ❑ Seeds/Plant components
    - ❑ Visible Parasites
  - ❑ Preserve
    - ❑ Parasite Identification

Basic Studies

Preliminaries Necessary to All Research

Exercise 1

Animal Identification: Learning Faces and Bodies

Exercise 2

The Ethogram: A Basic Behavior Inventory

Exercise 3

The Observing Schedule

Exercise 1

Animal Identification
Learning Faces and Bodies

A preliminary activity that is essential to the conduct of further research is the process of learning to identify individual animals or, if appropriate to the research protocol, to at least be able to identify individuals at the level of age and sex categories. It might be assumed that this is an elementary task, that everyone should be able to recognize a male and distinguish one from a female, and likewise, that it is simple to evaluate the age of an animal. In order to demonstrate that this may not necessarily be so, I suggest a visit to a zoological facility that keeps members of the genus *Ateles*, commonly known as spider monkeys. The usual response of students to the question of which is male and which is female in this genus will result in embarrassment. Without preliminary information to the effect that males have a retractile penis and the apparent "penis" observed is actually the enlarged clitoris of the female, the reverse of reality is usually perceived. Similarly in the past, practiced fieldworkers have made substantial errors in assessing the ages of subject animals since they were judging age on the basis of their own experiences and not upon criteria established over a long period (years) of familiarity with the subject species. It is wisest for the beginning observer to use very broad categories and to be cautious about estimates of age. Examination of known-age skulls and the degree of dental wear is a useful mechanism to enhance one's ability to estimate ages of primates.

Individuality: Recognizing That Not Everyone Is Alike

A key to beginning to recognize individuals is the acceptance that each and every organism produced from a sexual genetic union is a unique entity, different from all others. In addition it can be recognized that each individual pursues its own course through life and thus will have a further unique set of experiences as well as the accumulated evidences of minor and/or major traumatic injuries. All of these differentiating factors—genetic, ontogenetic, experiential—contribute to the identity of the each individual, and it is thus the responsibility of the observer to carefully seek out and record these distinctive features. It will generally be that these features are physical markings, variations in body structure, or scars, but it is not unusual for experienced workers to recognize familiar subject animals at inordinately long distances through some peculiarities of movement pattern or behavioral characteristics such as posture or carriage pattern. One of the easiest characteristics to discern is a limb, hand, or foot stiffness caused by a healed fracture.

In many captive care facilities and with some free-ranging populations it has become common practice to apply some form of tattooing to provide individual identification. A code system for facial tattoos is an occasional complement to Arabic numbers applied on a low-hair-density area such as the inner thigh or abdomen. Such codes like the one employed on the Arashiyama A and B troops (South Texas and Arashiyama Mountain, Kyoto respectively) are not intuitively obvious, so a researcher must be trained in their reading. In natural circumstances, various devices—collars, ear tags, ankle tags, and freeze branding—have been employed as identification tools, but have the major handicap of requiring that the animal be captured and sedated for their installation. In the end, most workers forego these extravagances and learn to identify individuals on natural characteristics. Since zoological facilities are unlikely to countenance manipulation of their charges, students will also have to learn individual natural features.

The Exercise

The purpose of this exercise is the development of skills at individual recognition and identification as well as to produce a set of primate identification cards sufficiently characterized that another observer could recognize the subjects from the material. It is estimated that this project should take between one and two hours to complete.

A. Select a subject group of primates. It is suggested that these should be a larger species such as a macaque, guenon, baboon, or an ape group. The group must consist of at least four animals. The reason for not suggesting a small primate such as a marmoset or squirrel monkey lies in their pattern and rapidity of movement, factors that would make it very difficult for a first-time observer to obtain appropriate information.

B. Examine the face and body of each individual for several minutes, paying attention to relative features that may serve to distinguish between individuals. Are there differences in sizes, overall dimensions; are there differences in sizes of specific regions or areas; are there differences in the hair coat (pelage) in terms of lengths on different regions, in colors; are there variations in skin (dermis) coloration or patterning? Are there any obvious scars or limb, hand, or foot damage? These type of distinctions will serve as the basis for the identification of individuals.

C. Using the identification sheets (either use copies of the examples here, or print out sufficient numbers of the correct form from the CD), fill in the appropriate data on the top of the sheet, and using the face schematics on the bottom, draw in the locations, sizes, and shapes of relevant pigmentation spots, tattoo marks, scars, other blemishes, or similar identifying features. It is undesirable to attempt to generate a perfect likeness of the subject; what is necessary are the key features that distinguish it from others, and, consequently, these can be very schematic (examine the sample).

D. The sheet has space for the observer's identity; the subject's identity (this may be the name used by the keepers, or it may be created by you as an aid to identification); group observed in; date of observation; age and sex evaluation; and, where known, the genealogical relationship of the subject. A further set of physical characteristics that may serve to separate this subject from others is also to be recorded. Take careful note of ear color and major notches (some forms such as lemur are often easier to identify by their ear notches—features produced during fights—than by other characteristics). The remainder of the sheet is available for other data.

A sample layout of the record sheet with an example is provided on the next page. The following pages provide blank versions for different kinds of primates, which can be copied to conduct the exercise. These sheets can also be printed as cards to be used in a permanent file. This design is based upon a card suggested in Appendix C of *Techniques for the Study of Primate Population Ecology* (National Research Council, 1981).

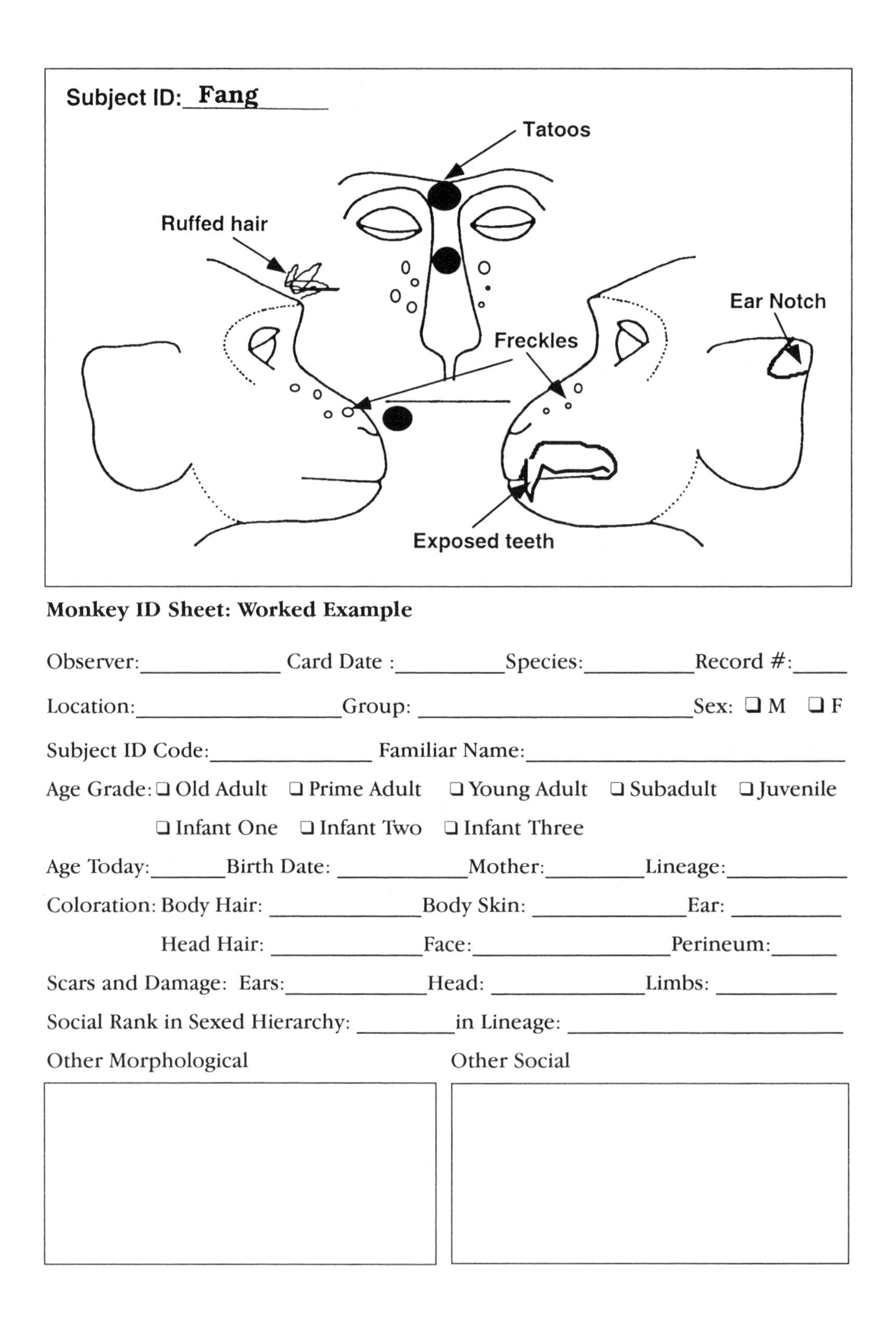

Monkey ID Sheet: Worked Example

Observer:____________ Card Date :__________ Species:__________ Record #:_____

Location:__________________ Group: ________________________ Sex: ❑ M ❑ F

Subject ID Code:______________ Familiar Name:_____________________________

Age Grade: ❑ Old Adult ❑ Prime Adult ❑ Young Adult ❑ Subadult ❑ Juvenile

❑ Infant One ❑ Infant Two ❑ Infant Three

Age Today:_______ Birth Date: ___________ Mother:_________ Lineage:___________

Coloration: Body Hair: ______________ Body Skin: ______________ Ear: __________

Head Hair: ______________ Face:__________________ Perineum:______

Scars and Damage: Ears:_____________ Head: ______________ Limbs: __________

Social Rank in Sexed Hierarchy: _________ in Lineage: _________________________

Other Morphological

Other Social

Subject ID:____________

Macaque Identification Sheet

Observer:____________ Card Date :_________ Species:_________ Record #:_____

Location:__________________ Group: _______________________ Sex: ❑ M ❑ F

Subject ID Code:______________ Familiar Name:___________________________

Age Grade: ❑ Old Adult ❑ Prime Adult ❑ Young Adult ❑ Subadult ❑ Juvenile

❑ Infant One ❑ Infant Two ❑ Infant Three

Age Today:______ Birth Date: ___________ Mother:________ Lineage:__________

Coloration: Body Hair: _____________ Body Skin: _____________ Ear: _________

Head Hair: _____________ Face:_________________ Perineum:______

Scars and Damage: Ears:____________ Head: _____________ Limbs: __________

Social Rank in Sexed Hierarchy: ________ in Lineage: _______________________

Other Morphological

Other Social

Subject ID:____________

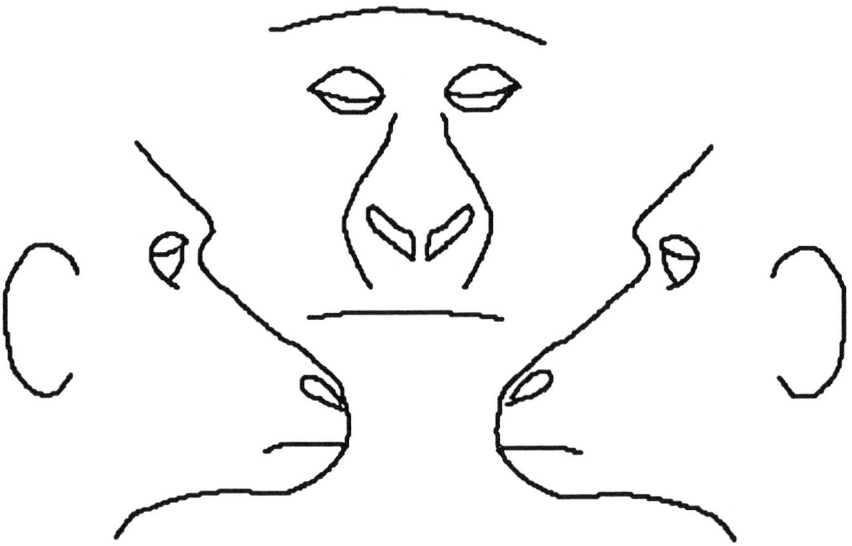

Gorilla Identification Sheet

Observer:____________ Card Date :_________Species:__________Record #:_____

Location:__________________Group: _________________________Sex: ❑ M ❑ F

Subject ID Code:_______________ Familiar Name:_____________________________

Age Grade: ❑ Old Adult ❑ Prime Adult ❑ Young Adult ❑ Subadult ❑ Juvenile

❑ Infant One ❑ Infant Two ❑ Infant Three

Age Today:_______Birth Date: ____________Mother:_________Lineage:___________

Coloration: Body Hair: ______________Body Skin: ______________Ear: __________

Head Hair: ______________Face:__________________Perineum:______

Scars and Damage: Ears:_____________Head: ______________Limbs: ___________

Social Rank in Sexed Hierarchy: _________in Lineage: __________________________

Other Morphological

Other Social

Subject ID:____________

Lemur Identification Sheet

Observer:____________ Card Date :_________Species:_________Record #:_____

Location:_________________Group: _______________________Sex: ❑ M ❑ F

Subject ID Code:______________ Familiar Name:_________________________

Age Grade: ❑ Old Adult ❑ Prime Adult ❑ Young Adult ❑ Subadult ❑ Juvenile

❑ Infant One ❑ Infant Two ❑ Infant Three

Age Today:______Birth Date: ___________Mother:________Lineage:__________

Coloration: Body Hair: _____________Body Skin: _____________Ear: _________

Head Hair: _____________Face:________________Perineum:_____

Scars and Damage: Ears:____________Head: _____________Limbs: __________

Social Rank in Sexed Hierarchy: ________in Lineage: _______________________

Other Morphological

Other Social

Subject ID:____________

Platyrrhine Identification Sheet

Observer:____________ Card Date :__________Species:__________Record #:_____

Location:__________________Group: _________________________Sex: ❑ M ❑ F

Subject ID Code:______________ Familiar Name:_____________________________

Age Grade: ❑ Old Adult ❑ Prime Adult ❑ Young Adult ❑ Subadult ❑ Juvenile

❑ Infant One ❑ Infant Two ❑ Infant Three

Age Today:_______Birth Date: ____________Mother:_________Lineage:___________

Coloration: Body Hair: ______________Body Skin: ______________Ear: __________

Head Hair: ______________Face:__________________Perineum:______

Scars and Damage: Ears:____________Head: ______________Limbs: ___________

Social Rank in Sexed Hierarchy: _________in Lineage: _________________________

Other Morphological

Other Social

Exercise 2

The Ethogram
A Basic Behavior Inventory

The concept of an "ethogram" is central to the study of animal behavior but is perhaps one of the more difficult to operationalize. An ethogram in its simplest sense is "merely" a catalog of an animal's behavioral repertoire, essentially a listing of the forms of behavior displayed by the animal. Unfortunately, complete ethograms do NOT exist for most primates in spite of many years of effort. Part of the problem is that it takes a very long period of observation—typically, thousands of hours—to produce a nearly complete catalog. However, it is generally expected that about 90 percent of a species' ethogram can be obtained in approximately one hundred hours of observation. Yet it is possible to obtain at least some idea of the potential behavioral range of a species from as little as one hour of observation. While such limited samplings will not produce a completely adequate listing of the species' ethogram, they will, if the animals are active, produce a record of a substantial portion of the animals' repertoire. The ethogram listing so produced, and the skills developed for doing so, will figure significantly in the other exercises in this book.

Behavior and Perception: The Influence of the Observer

It is important for beginning researchers to remind themselves that *a behavior record of the activities of subject animals is a mental construct of the observer and as such is subject to bias and distortion that may divorce the reality of the subjects' action from the perceived and recorded observations*. In a sense, many people see only what they want to see, even when they are trying to be objective. It is also important to recognize that this situation can become significantly worse if the observer is fatigued.

It is not just the vision system that may be influenced, observations may also depend upon auditory information, and the same processing pattern is applied by the observer to that data. In order to correct, partially, this set of problems, the threefold level of description for behavior presented in chapter 3 is an appropriate mechanism to aid in the control of perceptual biases.

The Ethogram

In the strict use and interpretation of the term, an ethogram implies only the molecular or molar level of behavior. However, the use of a cognitive hierarchy of behavior categorizations could be of great benefit to all primate observers. Still there are problems to be overcome. A major one is that in all cases, *it is desirable to formulate an ethogram with discrete categories of behavior with as few cases (preferably none!) of overlap as possible*. Unfortunately, the same atomic-level actions or units of behavior may be components of several distinct molecular units, and similarly of different integrated interactions. It is very often necessary to consider the context in which the behavior occurs before one dares to attempt interpretation. The construction of ethograms is a project area littered with traps and pitfalls. With these encouragements and advice, we now come to . . .

The Exercise

The purpose of this exercise is to learn to perceive and categorize the behavior of primate subjects in a captive, controlled environment. Secondarily, the exercise provides practice at constructing definitions, an activity that is central to any study of behavior.

The observational and exercise procedures are as follows:

A. Select *a single individual of a single species* as the subject. The subject should be moderately active (*too inactive and there will not be much activity or variation of behavior to observe, too active and it may be impossible to keep the records without losing data*).

B. Observe the subject for a period of two or three hours (your instructor may specify more or less time for this exercise). It may be advisable to take a half hour to familiarize yourself with the subject group and their activity patterns before selecting a subject and beginning the formal study.

C. Record—as molecular behaviors—each of the activities of the subject during the period. Once a particular activity is seen, it need not be repeated in the record unless it reflects a change from active to passive or vice versa, or is changed in form.

D. Carefully watch and note the details of each of the activities and try to recognize the identifying or diagnostic features of each, that is, those features that make behaviors distinguishable from each other.

E. Write up definitions based on these detailed descriptions as *formal definitions* of each behavior unit. *Beware of the cardinal rule of definition: A term may not be described using the term being defined.* It is NOT acceptable to define sleeping as "the activity of sleeping," but "remaining in a relaxed posture with closed eyes, regular breathing, and apparent unconsciousness" is *one* satisfactory definition. These units are perhaps best described in the form of an abbreviated atomic description of the actions that constitute the unit.

The results can be presented as a short report or as a labeled list. Remember to identify yourself adequately. A basic form is included on the CD and, at the discretion of the instructor, may be used for completion of this exercise.

How many different behaviors were you able to find in your study? It would take a very active subject and very close observation to find more than thirty distinct behaviors in such an initial study, but you should be aware that an ethogram for baboons (*Papio cynocephalus*) produced by Coelho and Bramblett (1990) counts almost two hundred distinct behaviors at their *molecular* level.

Exercise 3

The Observing Schedule

An observing schedule is an important preliminary to the conduct of any project that involves controlled sampling. Several of the techniques in the exercise set do not require a schedule since there is no requirement for equality of sample sizes or "stationarity" in the data for the statistics used in their analyses. (Stationarity "means that the sequential structure of the data is the same independent of where in the sequence we begin" [Bakeman and Gottmann 1997: 138]. Thus, the first half of the data will be similar to the second half, and both will be close to the characteristics of the whole set.) However, other protocols do require either or both of these features for a valid analysis to be produced. Interval sampling can fall into both categories: instantaneous scan sampling, as repeated slices of behavioral time in a group, does not, but focal time sampling, as repeated slices of behavioral time for one individual at a time, does require scheduling. The most frequently employed system, continuous recording of individual behaviors for fixed periods—focal animal sampling—is the protocol most in need of schedules.

As noted in chapter 2, there are two forms of observing schedule, fixed and random. Consequently, the exercise involves production of both types of schedule.

THE EXERCISE

Fixed Format Schedule

The student is first directed to the section on fixed format scheduling in chapter 2, and will then set up a schedule of observing for the following project.

A population of spider monkeys, *Ateles paniscus*, resides in a large outdoor enclosure 20 by 20 meters and 10 meters high, at the local zoological park. The group consists of one adult male, three adult females of differing ages who are all descendants of a founder female two generations in the past, four juveniles (three female, one male), and two older infants. The facility is open all year round from dawn to dusk, and the variation in day length is limited (fourteen hours at midsummer, ten hours at midwinter).

A study of fourteen days total duration is planned, with approximately one week scheduled for midsummer and another for midwinter. The observing protocol employed is to be half-hour (thirty-minute) focal animal samples. The observer feels that a five-minute rest after two samples is appropriate, and that a six-hour, twelve-sample working day is all that is possible. The sample sizes must be equal for all animals, and for each of the study periods. Questions appropriate to this task are:

- How many hours are available to the study? How are they distributed? How will the working days be distributed between the two study periods?
- What is the influence of the rest periods on the sampling schedule? What alterations would adjust the schedule to make it more uniform? Is it necessary?

The tasks for this exercise are to:

- Construct schedules (a row and column table with codes for individuals is appropriate) for the summer and winter study periods.
- Construct schedules with your choice of an alteration in the observing protocol in agreement with question 2 above.
- Construct schedules for two observers working on the project and attempt to maximize the amount of observation time.

You should end with six schedule tables. A blank sample schedule sheet is included on the CD.

Random Sampling Schedule

The student is first directed to the section on random scheduling in chapter 2, and will then set up a schedule of observing for the following project.

A population of thirty-six identified individual olive baboons, *Papio cynocephalus anubis*, is to be the subject of a intensive study in the woodland savanna near the edge of the Albertine Rift in western Uganda. The observing protocol has been decided as fifteen-minute focal animal samples followed by a five-minute rest and search period, with the observer putting in a twelve-hour field day. The study is intended to be one full year in duration.

The tasks for this exercise are to:

- Using either a set of strips numbered 1 through 36 or a pair of cubic dice (see below for instructions on how to operate this process), construct an observing sequence for a six-day working week.

- Total the numbers of samples for each animal in the study population in order to see how often each individual is studied during the week. Is the distribution equitable? How many subjects were not to be observed at all during this particular week?

Random sample generation using dice is not as simple as the use of numbered slips. Since the most common die is the cube with numbers from 1 to 6, it requires a stratified process for any larger group. The simplest technique is to divide the entire group into six parts, and for very large groups each of these may be divided into further six parts. At that point it is possible to start rolling the dice. In the case above, with thirty-six individuals, two dice can be used. The first die selects the partition and the second selects the individual within the partition (to keep it clear the dies should be of differing colors). Thus the first die, if it comes up with a 3 selects the third partition, that is, individuals who are numbered 19 through 25, and if the second die also turns up 3, the third individual in the set is selected—number 21. The die readings themselves do not indicate individuals unless there is equivalence between the number of subjects and the number of faces on the die. Role-playing dies with up to twenty faces exist and could be used in place of the stratified process, but with higher numbers one is again faced with the need for stratification of the process. Since each selection of a partition depends upon the random turn of the die, the end result is a true random selection of a subject.

Remember that a reserve list should also be generated to cover cases where the target cannot be located for that sample period.

Group Comparative Studies
Appreciating Differences Between Species

Exercise 4
The Behavior Profile: A Comparative Study

Exercise 5
Time Budget: A State Behavior Duration Exercise

Exercise 4

The Behavior Profile
A Comparative Study

The behavior profile is a study technique very appropriate to the usual settings of a zoo environment and is designed to produce a direct comparison between two or more species of animals. As outlined below, the technique is simple but effective; however, the concept and construction of an ethogram as used in one of the previous exercises is an important prerequisite. In one sense there is a degree of similarity between this exercise and exercise 5 on time budgets; however, the purpose of this one is oriented toward observing the differences in the *frequencies* and *relative frequencies* of behaviors, not the amounts of time spent at each.

The behavior profile exercise requires that the observer begin with the construction of an arbitrary ethogram, alternatively the ethogram from exercise 2 can serve as the basis for this project. However, the ethogram should be quite general and limited to a few categories, certainly no more than eight or ten. It is important to use categories of behavior that the student feels will describe the activities of the subjects in each of the comparative groups. These behavior categories are to be constructed and defined as states for this exercise, even though they will be treated as events in the data collection and analysis. An appropriate example for an ethogram would be something like this list: locomoting, resting, feeding, scanning environment, aggressive social interaction, peaceful social interaction, grooming, sexual activity. Definitions of these are left up to the observer, and these are not an exclusive or required set; construction of other sets is encouraged.

Each initiation of these states, that is, each change of behavior, during a fixed observation period is recorded. In a sense, this is a recording method very close to the "all occurrences" sampling technique referred to by Altmann (1974). For the analysis, frequencies of each behavior in each species are calculated. Note that this pattern is one of the variations in the treatment of behavior states; they either can be handled as durational data or counted as occurrences—just as if they were events. This type of treatment loses a great deal of information; however, no more than in some other data-gathering methods.

THE EXERCISE

The observational procedure is as follows:

A. Select four different species of primates upon which to conduct the study; the four groups should have comparable sizes, but that is not a rigid requirement. A difference of one or two individuals should not be significant in a larger group but definitely will be if small groups are compared with large groups.

B. Use a watch, or stopwatch, to time a fixed observational period for each species; fifteen, thirty, or sixty minutes for each is suggested as appropriate, or as selected by the instructor.

C. Record all occurrences of the onset or beginning of each of the behaviors taking place within the enclosure; this means that all members of the group are being scanned, but *attention is focused on behaviors*, and the changes of behavior, not on individual animals. *Note:* Behaviors that are continuing are counted only once.

D. Repeat the fixed observation period for each species group.

   Observational Hint:
   The simplest method is to record a "tick" in the appropriate check-sheet column at the onset of the specified behavior category, irrespective of which animal of the group does it. If two animals engage in a fight, then two ticks are recorded; if three animals begin resting over five minutes, then three ticks go into the resting category. Remember that the focus of attention is on the activities of the group as a whole, NOT on any single individual. Once the data have been collected from the four species for the same periods of time (periods of fifteen, thirty, or sixty minutes have been suggested as appropriate, but other periods may be used), add up the counts for each category and also the total for each species.

E. Prepare a brief report. This report should include a written component summarizing the method, results and a discussion, and at minimum a data table showing the frequencies, relative frequency (calculation *B* on p. 55), of behaviors and rates in either occurrences per minute or per hour, whichever provides the most readable values. A column or bar graph of either relative frequencies or hourly rates must also be included.

To aid in getting started, a sample check sheet, using the ethogram list suggested above, is provided as a model and is available on the CD. Students are encouraged to create their own rather than to rely explicitly upon this sample.

Check Sheet For Behavioral Profile

Observer:______________________________ Date:____________ Time:_________

Sample Duration:______________ Location:_________________________________

Species 1:____________________________ Species 2:_________________________

Species 3:____________________________ Species 4:_________________________

Behavior	Species 1	Species 2	Species 3	Species 4
Locomoting				
Resting				
Feeding				
Scanning				
Aggressive				
Peaceful				
Grooming				
Sexual Act				

Exercise 5

Time Budget
A State Behavior Duration Exercise

The time budget exercise is related to one of the important types of research conducted on wild primates during the past few decades. The manner in which a primate, individually or as part of a group, divides up its waking hours is considered to represent an important aspect of its ecological adaptation. This is known as time budgeting and represents the priority of resource utilization in the activities of making a living. Time budgeting has also been severely criticized by Hilary Box (1984) on the basis that it is

> difficult to obtain accurate measures of time budgets. It is difficult to compare directly results obtained by different methods of estimation. Apart from methodological constraints, studies which have used the same methods of estimation have shown that different populations of the same species have varying patterns of activities in different ecological conditions. (Box, 1984:15)

While Box is correct that the problem is complex and that it is extremely difficult to compare across populations, let alone species, the basic concept of a time budget remains valid, particularly for a single population in a specific environment over a relatively short period of time. This suggests that, when treated with care and understanding of the constraints on the protocol, there is sufficient internal consistency to make time budgets useful. Recent examples of more extensive time budget studies than this exercise requires can be found in Defler (1995), Menon and Poirier (1996), and Kurup and Kumar (1993).

The amount of time that a primate spends in feeding and foraging activities is governed by a number of factors that are to some extent species specific, and to some extent are related to the ecological situation. These factors include such things as:

- Absolute size—a larger primate obviously needs more food than a smaller one, but that need is scaled allometrically such that larger forms actually display an increase in energy efficiency.
- The type of food that the species feeds upon—animal flesh has the highest concentration of available necessary proteins but is energetically very costly to obtain; plant starches are easily locatable but can be energetically expensive to digest; and fruits are intermediate to both but are only sporadically available.
- The distribution of the required foods within the environment—some are patchily distributed in clusters while others are uniformly and/or sparsely scattered across the environment, thus dictating different searching strategies.

If a primate species feeds on fruits, its digestive processes tend to run fast and to extract relatively fewer but higher energy nutrients from foods since extraction is a time-dependent process. Conversely, leaf eaters must spend a great deal of time processing their foods since they depend in some measure on symbiotic bacteria within their digestive tracts to extract the low amounts of energy in the cellulose and hemicellulose portions of plant structures. As a consequence of this difference, fruit eaters tend to be much more active and engage in foraging to a greater extent than leaf eaters, who spend a greater amount of their time digesting. An appropriate analogy is the difference between a banquet where the guests sit in one place eating and digesting, and a buffet where they are continually returning to sample the available delicacies, interspersing periods of feeding and moving.

In similar fashion, the locomotor activity patterns of primates are to a great extent controlled by the amounts of available energy, the ambient temperatures, and the immediate social environment. This varies for every individual within the society; consequently, the time budget for one individual is rarely coincident with that for any other individual.

Often, one of the most surprising findings for students of primate behavior is the amount of time that the study subjects spend sitting around apparently doing nothing! To some extent this surprise is a product of the video and films utilized to show primate behavior. These are carefully selected and edited from the available footage to illustrate the *active* behavior of the species, and, hence, they tend to produce a false impression of the amounts of time spent by the subjects at various activities. This is perfectly understandable; no one would sit still and watch three or four hours of film of a primate sleeping, yet fieldworkers often find themselves performing that very type of activity. I once spent five and one-half hours in a Mexican rain forest with a solitary subadult male howler (*Alouatta palliata mexicana*), who slept the whole time and provided only four records of change in posture. The cautionary message is: Do not assume too much about the activity pattern of individual primates, and especially do not rely upon media-derived impressions.

THE EXERCISE

This exercise takes the observer through the activity of conducting a brief time-budget study of *at least three hours' duration*. (To complete properly, such a basic study would take a minimum of one full day for each member of the study group.) Instructors may make substantial changes in the hours required for this project.

The technique of conducting the study is straightforward. Basically, an ethogram of behaviorial state categories that reflect gross activity and energetic cycles is constructed. The traditional set of behavior categories used, and advocated by many researchers for comparative use, consists of Rest, Move, Feed, Social, and Other. This list is sometimes reduced even further to Rest, Move, Feed, and Other but may be expanded for specific purposes (as will be seen below). On a record sheet, record the onset (start times) of all activities as they occur. All of the behavior categories must be defined in such a way that they are mutually exclusive, thus avoiding the possibility of two activities being recorded simultaneously. This exercise can be done with a small group (three to five individuals), using a check sheet set up with columns for each subject and using a simple code (R, M, F, S, O or R, M, F, O) and the time of the activity start. This process can be very tiring and is probably best broken up into fifteen-, thirty-, or sixty-minute sessions. An alternative methodology (employed in form B of the exercise), and one that is essential for a group larger than three to five, would be to focus upon individual animals either for fixed-duration samples at specific hours on equivalent days, or through the development of a random sampling schedule. These options would mean a greatly expanded amount of observational time (three hours times the number of group members). In either case a variation of the check sheet suggested in the scan sampling exercise (exercise 6) is available at the end of this exercise and can be found on the CD.

Since the onset of one activity will also be the offset (end time) of the previous behavior, it is not generally necessary to record offsets (a very good reason to define the categories as mutually exclusive). The categories used should be capable of reflecting as near to 100 percent of the time spent observing as is possible. At the end of the observations it is a simple procedure to calculate the percentage of time spent by the individual subjects (or by each age-sex category) at each of the activities. This reflects the time budget of the sample population. The formula can be stated as:

$$\frac{\text{Total Time in Activity}}{\text{Total Sample Time}} \times 100 = \text{percentage Time}$$

The results should be presented as a standard, written scientific report, although a brief report format may be specified by the instructor. It is advisable to conduct some library research and include a list of references used in either case.

Exercise Form A

This form of the exercise requires three hours' minimum study time and the use of a check sheet for parallel observation of three to five individuals. The ethogram used should be either the RMFSO or the RMFO versions. The start time and code for each behavior change is noted in the column for each subject, and the sample should continue for fifteen, thirty, or sixty minutes. Breaks should be taken after each sample session.

Exercise Form A2

Form A2 of the exercise is the same as Form A but employing the focal animal sampling procedure. This means that observation focuses upon one individual subject at a time for a period of time (same sample durations as above). The sample size (number of individuals) for this version is expected to be larger than in Form A.

Exercise Form B

Form B of the exercise requires the use of a more extensive ethogram. The behavior listing can be as extensive as student and instructor agree upon but must consist of only state behaviors—all event behaviors can be lumped under the category of "other." The ethogram list with definitions should be presented as part of the report. As in Form A, parallel observation of three to five individuals for three hours is required.

Exercise Form B2

Similar to Form A2, the Form B2 exercise is based on B. The requirement is for an extended listing of state behaviors (with presentation of the ethogram definitions) but employing focal animal sampling procedures. The group used is expected to be somewhat larger, and, consequently, the time involved in the study will be more extensive.

THE REPORT

Irrespective of which form of the exercise or whether a brief or full report is required, all papers should include the following components:

- the date of your study—not the date of writing it up
- the duration of the study, or if it is broken up into random or scheduled samples on individuals, the duration of each sample and the total sample time
- species name and the figures for the group composition
- the behavior categories used and their definitions
- presentation of a summarization table of your data collection (This should include the durations of sample time for each age-sex class studied, and the percentage of that class's total time, for each category of behavior.)
- a graphic presentation of % Time for some subset of age or sex or age-sex categories
- a discussion of the data presented
- presentation of any conclusions that you have reached about the time budget of the species that you have studied (This is the place to confirm or deny any hypotheses or conclusions that you may have found in your *readings*. But be sure to take some cognizance of the differences between natural and captive conditions.)

Check Sheet For Time Budget

Observer:____________________________________Date:____________________

Subject Group:_________________________________# in group:______________

Start Time:______________________Group Composition:_____________________

Time	Subject 1	Subject 2	Subject 3	Subject 4	Subject 5

Major Modes of Observational Methods

Learning to Sample Effectively

Exercise 6

Scan Sampling: Instantaneous Scan Sampling of States of Behavior/Interval Sampling

Exercise 7

One/Zero Sampling (sometimes called Hansen Frequencies)

Exercise 8

Focal Time Sampling:
State Behavior Interval Sampling Procedure

Exercise 9

Focal Animal Sampling: Intensive Study of Behavior

Exercise 10

All Occurrences Sampling:
Collecting Specified Forms of Behavior

Exercise 11

Matched Control Sampling:
Studying Reconciliation Behavior

Exercise 6

Scan Sampling
Instantaneous Scan Sampling of States of Behavior/ Interval Sampling

The process of scan sampling, often called interval sampling, can be considered as analogous to the repeated sweeps of an air traffic controller's radar screen. In much the same way, a sweep across the members of the subject group is made, and an instantaneous observation of the state of behavior displayed by each subject individual is recorded. *Theoretically* this sweep is conducted in zero time, and the data collected is then considered to reflect the behavior of the group at a precise slice in time. *Practically*, however, it does take a few seconds to complete the scan, but since it is *assumed* to be instantaneous, observers should attempt to perform a scan as quickly as possible and yet take sufficient time to retain accuracy. This technique is most frequently oriented to describing group and subgroup (e.g., age-sex class) behavior rather than individual behavior. This makes the assumption that the group is a large one. In a small group, such as is typical of zoological collections, it can be functional at the individual level as well, on condition that the individuals are clearly recognizable and can always be identified during the scan.

As a research method it is particularly suited for dealing with activity-cycle studies where it can be used to generate an estimate of "percentage of time engaged in specific activities." It is not a suitable technique for the study of interindividual interactions, as these tend to occur in sequences that cannot be adequately sampled. It also misses nearly all of the short-duration (event) and infrequent behavior. One very appropriate use of scan sampling with captive primates is to determine the pattern of preferential usage of various areas in the cage (see exercise 20).

Required Conditions for Scan Sampling

The use of scan sampling techniques must be appropriate to the research design of the study and the particular type of data to be collected. *The behavior to be observed must be of the* "state" *form*, the individual members of the group must be identifiable to at least age and sex class (in small groups it should be to individual level), and the behavior categories must be familiar to and easily recognized by the observer. In cases where these conditions are not consistently present, as when a particular behavior is not perceptible to the observer immediately, a degree of bias is developed that can distort the results.

To minimize problems, the scans should be performed in as consistent a manner as possible throughout the study.

Since the most common end use of scan sampling is to produce a percentage estimate (either frequency or duration) of activity or location, the interval periods of the scans are very important. The closer the scans are to each other, the closer the estimate can come to achieving a measure of the "real" time-use pattern. In order to have any reasonable degree of confidence in the validity of this method for real-time estimates, it is necessary that the scans be no more than ten seconds apart, and preferably only two or five seconds. There are a number of intervening factors such as timing control, the type of activity, the size of the group, and the difficulty of behavioral unit recognition or individual subject identification, which will dictate the functional minimum periods between scans. Researchers often use intervals of fifteen, twenty, or thirty seconds between scans, and a common accessory is a "beeper box" to provide precise control over the scanning rate. (A schematic for a simple beeper can be found as appendix 4 in Martin and Bateson [1993], an older version is in appendix 2 of the first edition.) Many modern digital watches have "pacer" functions that can be set to operate in the same fashion (several Timex and Casio sport watches have this capability).

Correction of the Out-of-Sight Problem

As mentioned in chapter 2, one of the significant problems that can develop in a short study is the out-of-sight situation. This is particularly important in this exercise since out-of-sight records for particular individuals or age-sex classes will significantly bias the results. This biasing will be particularly evident due to the short time of the study—it would tend to even out and become less of a biasing factor if a hundred or more hours of sampling were being done. In order to deal with the out-of-sight problem in this exercise, the observer has a set of options available. The first is obvious, but rather disheartening: Throw out the sample as soon as an animal goes out of sight and begin again. A partial solution that avoids this scenario would be to conduct the exercise on a warm, sunny afternoon rather than a cool, damp or rainy morning—there would be a greater probability of completing the exercise with all animals visible. The second option is to suspend scans while any animal is out of sight and then continue once all subjects are visible again. This would allow the collection of a complete sample, but as discussed in chapter 2, may introduce unknown and uncontrollable biases into the data. It is, however, an acceptable choice in this exercise. The third option involves two stages. The data must be collected in the standard format, but if any individual animal is absent from view for more than one-third of the samples, the sample must be redone. The second stage involves a correction ("a fudge factor" to some) that is applied to the data set. This correction assumes that at least two-thirds of the sample have all subjects present, and the calculations are based only on those scans. Effectively, one removes the complete scan in which an animal is out of sight. If this means the exercise ends with 143 valid scans instead of 160, the calculations of rate and percentage will be influenced if uncorrected. For example, if feeding occurs 34 times in the whole sample, the calculations would be:

for 160 scans   34/160 = .2125 as the rate per scan
34/160 x 100 = 21.25% (percent of scans on which feeding occurred)

for 143 scans   34/143 = .2378 as the rate per scan
34/143 x 100 = 23.78% (percent of scans on which feeding occurred)

To calculate the rate per hour or the rate per minute in the same situations, the following calculations would take place:

for 160 scans or 40 minutes
34/40 = .85 per minute
34/40 x 60 = 51 per hour

for 143 scans or 35.75 minutes
34/35.75 = .951 per minute
34/35.75 x 60 = 57.06 per hour

The use of percentages and corrected calculations for scan sampling is the most appropriate mechanism to improve the accuracy of the data summarizations.

Discrepancies in the sex ratio, the numbers of males and females in the group, will also have some biasing effect on the overall group frequencies and rates. If a comparison between the behaviors of males and females is attempted on the basis of simply adding the scans, it will produce significant differentiation between the sexes. However, if the data are averaged for age-sex classes, and especially if converted to percentages, the result will be more acceptable. But for this exercise those problems can be ignored.

The Exercise

This exercise involves conduct of a limited study of a group of primates using the scan sampling technique.

The procedure is as follows:

A. Select a social group of at least four animals. It would be best to use an active group of monkeys or prosimians such as squirrel monkeys, ringtailed lemurs, spider monkeys, or some species of macaque.

B. Spend sufficient time in familiarizing yourself with all of the individual members of the group to enable you to identify each one quickly and accurately (perhaps a review involving exercise 1 would be appropriate).

C. Establish that your ethogram is suitable for the subject group; if it is not, then some modification is warranted.

D. With check sheets (a sample is provided on the following page and on the CD) and a watch or stopwatch, record forty scans, employing intervals of fifteen seconds. Be sure that you include every member of the group in each scan. This will take ten minutes; then take a few minutes as a break while preparing for the next sample.*

E. Repeat this sequence four times over a period of one hour; that is, four 10-minute record sessions, which will total up to 160 scans.

F. Calculate the *frequency*, the *rate*, and the *percentage* for each of the behaviors observed for the group as a whole, and then for each age-sex class represented in the group. Finally, prepare your report.

The check sheet on the following page is laid out to handle scan sampling of a group of five individuals. Note that the sheet would have to be expanded to handle larger groups.

**Note: One of the most important problems for observers of animal behavior is "observer fatigue," which can rapidly lead to a degeneration in the quality of data collected. To prevent this occurrence, regular rest breaks are a necessity.*

Check Sheet For Scan Sampling

Observer:______________________________ Date:__________________

Subject Group:__________________________ # in group:____________

Start Time:__________________ Group Composition:__________________

Time	Subject 1	Subject 2	Subject 3	Subject 4	Subject 5

Exercise 7

One/Zero Sampling (Sometimes Called Hansen Frequencies)

J. Altmann (1974) *does not recommend the process of one/zero sampling*, as the protocol does not produce data that is either frequency or duration. However, she has noted that if a biologically relevant behavior that occurs infrequently enough that a simple presence/absence score is useful, then one/zero sampling might be applicable (J. Altmann, personal communication, August 1994). Several other researchers do accept the usefulness of this procedure; in particular, the late Ray Rhine and his collaborators have argued that the data is neither frequency nor duration but reflects contributions from both and, consequently, is a reliable and objective measure of behavior (Kraemer, 1979; Rhine and Ender, 1983; Rhine and Flanigan, 1978; Rhine and Linvillr, 1980; Rhine, Norton, Wynn, and Wynn, 1985). Bernstein (1991) points out, in a very reasoned comparison, that this mixed condition is the core basis of objections to the procedure.

> Critics indicate that the relative contribution of frequencies and duration is a function of the time unit selected. They prefer data on true frequencies or durations as a direct description of animal behavior and reject an abstraction based on a non-specified combination of both. (p. 727)

Bernstein goes on to state, in parallel with Altmann, that a biologically appropriate circumstance might lead to the use of one/zero sampling, when "responses occur in non-independent flurries, where the bout, and not each individual act, is of interest" (Bernstein 1991:727). He suggests that situations such as bouts of play behavior, or calculating the probability of receiving aggression during a specific time period, might be such situations.

Kraemer (1979) has also discussed the use of one/zero sampling, but suggests that alternatives to one/zero sampling should be used whenever possible.

With all these caveats in mind, and a quick check back to Figure 2.5 in chapter 2 to remind ourselves of the difference between sample intervals and sample points, the observational method can be outlined, and the exercise can be set up.

One/zero sampling involves the establishment of a set of observational sample intervals. The intervals may be of any length, though the practical limits are probably twenty-four hours at the upper end, and two seconds at the lower. For the practical purposes of this exercise, and to show the limitations of the method, periods of ten, fifteen, thirty, or sixty seconds are suggested, and the period may be explicitly required by the instructor. Timing using a watch or stopwatch may be difficult, but many digital watches now have the capacity to provide an audible beep at intervals selected by the operator, or a beeper box might be available. The recording procedure can be done on blank paper, but a prepared check sheet is more useful. Once observation begins, the observer keeps track of the activities of the subject or subjects and at the end of each sample period places a tick in the box for the behaviors seen, and a 0 in the box for the behaviors not seen. No count is made, as the question is "did the behavior occur?" not "how many times did the behavior occur?" An alternative mode of operation is to place the tick mark in the appropriate box as soon as the behavior occurs. This has the advantage of eliminating a memory search ("did it or didn't it occur?") at the end of the period, and in immediately removing that behavior from further consideration. As with some other methods, it is possible to perform one/zero sampling on a group or on individuals chosen as subjects one at a time. The results from these two variants are not comparable. The only score or measure that is derived from one/zero sampling is a proportion. The proportion of sample periods in which the behavior occurred is calculated by simply counting the number of periods with a tick and dividing by the total number of periods used for observation. Going beyond this simple calculation is not statistically appropriate.

THE EXERCISE

This exercise involves conduct of a limited study of a group of primates using the one/zero sampling technique.

The procedure is as follows:

A. Select a social group of at least four animals. It would be best to use an active group of monkeys or prosimians such as squirrel monkeys, ringtailed lemurs, spider monkeys, or some species of macaque. The instructor will direct you to use either a group or individual sampling schedule.

B. Spend sufficient time familiarizing yourself with the group to establish that your ethogram is suitable for the subject group; if it is not, then some modification is warranted.

C. With selection of an appropriate sample period—five, ten, fifteen, thirty, or sixty seconds as suggested or required by the instructor—set up a check sheet for the project.

D. With check sheets (a sample is provided on the following pages, and on the CD) and a watch, stopwatch, or beeper box, record forty sample periods. This will take ten minutes if fifteen-second periods are used, forty minutes if sixty-second periods are selected; then take a break while preparing for the next sample.

E. Repeat this sequence four times to total 160 sample periods.

F. Calculate the proportion for each of the behaviors observed for the group as a whole, or for each age-sex class represented in the group if individual one/zero sampling is to be used. Prepare a data table and a graph showing the variations in proportions found. Finally, prepare your report.

The sample check sheet on the following pages is laid out for one/zero sampling with a maximum ethogram of ten behaviors and forty sample periods. As always, it can also be found on the CD.

Check Sheet for One/zero Sampling

Observer:________________________________ Date:______________________

Subject Group:____________________________ # in group:_________________

Start Time:_______________ Group Composition:________________________										
Behaviors										
T										
1										
2										
3										
4										
5										
6										
7										
8										
9										
10										
11										
12										
13										
14										
15										
16										
17										
18										
19										
20										
21										
22										
23										
24										
25										
26										
27										
28										
29										
30										
31										
32										
33										
34										
35										
36										
37										
38										
39										
40										

Exercise 8

Focal Time Sampling
State Behavior Interval Sampling Procedure

The procedure called focal time sampling is a specialized technique combining some of the features of both focal animal sampling and scan sampling. It was introduced by Baulu and Redmond in 1978 as an alternative and addition to these two methods of sampling. Focal time sampling is oriented toward the acquisition of state behaviors from a single subject, the focal subject, at specified intervals over a fixed sample period. This technique is one variation of interval sampling but has the added values of individual identification and, when the interval is short, an approximation to real time. It is also possible and appropriate to add regular, nonbehavioral variables into the procedure for ecologically orientated studies. That is, at each interval, the behavior and other variables such as substrate used, temperature, nearest neighbor, and so forth, can be recorded.

In practice, as for focal animal sampling, an observing schedule of focal sample periods has to be preestablished, and the sample period is, in each case, devoted to recording only *the activities of a single subject plus any select additional variables. It also means that like scan sampling, the records are only of state behavior, and they are taken at predetermined intervals during the sample period, using the same assumption of instantaneous recording as in that technique.*

As a consequence, the data records obtained from this method consist of a series of states for the subject taken at intervals, although it is likely that an observer will take data on several aspects of the behavioral state and, perhaps, environmental factors at each interval. This type of research method requires a well-conceptualized research design and typically will employ a distinctive method of analysis.

Method and Design Requirements

The method of conduct of focal time sampling is extremely straight forward, but is intimately tied to research designs that are oriented toward two particular kinds of analyses—correlations and sequential procedures. As in all observational research it is necessary to have an appropriate ethogram for use, and, in this case, the exercise of constructing an observing schedule is a vital prerequisite. The observing procedure during each sample is then quite simple. The observer maintains visual contact with the selected subject animal, and from the beginning of the sample period records (through the use of check sheets or perhaps with an electronic device, such as a data logger) the behavioral state of the subject at fixed intervals. The intervals and duration of the sample periods are fixed by the observer to be appropriate for the research design objectives. These might vary from ten-minute samples with observations taken at intervals of ten seconds (yielding sixty records per sample), to sixty-minute samples with observations at intervals of two minutes (yielding only thirty records per sample), to anything that the observer considers appropriate for the species and research paradigm. Ancillary data may also be collected with each observation, such as posture of the subject, various orientations to social and/or environmental factors, climatic factors, ecological factors, and so forth, but the observer must be careful to record these as quickly and concisely as possible in order not to violate the assumption that the whole set represents a single slice in time, that all of the data are acquired at the same instant.

Research Designs

The research designs employable with focal time sampling are those that are predisposed to the types of statistical analyses that were mentioned earlier—correlations and sequential statistics. Thus, the most appropriate kinds of designs are those in which several distinct and independent variables are collected simultaneously with a dependent observed variable for which the independent variables are hypothesized to be causal. As an example, in a study using this method, a dependent variable—the posture of the subject—was recorded along with a set of independent factors—subject's activity, amount of exposure and orientations to wind and sunlight, altitude above the ground level, the material substrate on which the subject was positioned, and the ambient temperature in the immediate vicinity of the subject. The operating hypotheses were that one or more of the independent factors exerted an influence in inducing the subject to assume various postures that represent changes in the amount of surface area exposed, which in turn indicates aspects of the heat exchange pattern (the results may be viewed in Paterson, 1992). Many variations on the design of research projects are possible, and the student is encouraged to be creative in formulation and critical in evaluation of projects.

ANALYSIS TECHNIQUES

The data set from a focal time sampling study may be treated with standard descriptive statistics in order to yield frequency and relative frequency per sample, or the data may be converted to yield rates, in particular bout rates; when the intervals between the samples are short enough (it was suggested by Baulu and Redmond that ten seconds is usable for this, and other analyses have confirmed its suitability), mean durations per bout and per hour and percent of time can also be derived. Thus, many of the same table and graph formats for presentation of results are appropriate as for the other methods of observation. Those students with more advanced statistical training may wish to consider application of multiple regression, Markov chain and lag sequential analysis, or the Lisrel procedures, to analysis of their data.

THE EXERCISE

The purpose of this exercise is to provide practice in using the focal time sampling technique (FTS) and preobservation preparatory work in the study of primate behavior. The exercise is to be conducted as follows:

A. Prepare an appropriate *ethogram* and *an observing schedule* for study of a group of four or more animals using FTS for ten-minute samples with ten-second intervals (sixty records per sample). Each animal should be studied several times (the total time involved in the study depends upon the number of animals in the group and the number of repeat cycles specified by the instructor). An example of an appropriate study question is: are there age or sex differences found in the behaviors within the study group?

B. Prepare check sheets for the study—the example for the scan sampling exercise can be utilized, or the version on pages 148 and 149 (a single page version is on the CD) may be more suitable.

C. Using a watch, stopwatch, or beeper box to control the time intervals, conduct the research phase.

D. Perform the analysis of the data and include the calculations of mean rate per individual (MRI) and mean duration per individual (MDI) as found in chapter 4. These calculations apply to the entire group or to comparative classes such as male/female, old/young, and so forth, but may *not* be applied to individuals. Another useful calculation would be the mean duration per bout (MDB), which is focused upon differences in the behaviors of the ethogram, but could also be split along age and sex lines. It is important to remember the difference between a bout and a record. It is important to provide the total number of scans, the total time involved in the study, and the number of scans per individual in the results.

E. The results may be written up in either a brief or full scientific report as required by the instructor.

An Advanced Version of the Focal Time Sampling Exercise

For the student with a knowledge and understanding of inferential statistics, in particular, of correlation, regression analyses, or sequential analysis, a suggested variation on the exercise is as follows.

A. Set up a research design involving the study of a particular aspect of behavior and some structural, locational, climatic, or other easily measured environmental factor. This will involve the construction of an appropriate set of hypotheses about the relationship, and designing the protocol to use FTS in a manner that will provide data to test the hypotheses. If the intention is to focus upon sequential analyses, this is irrelevant and only a behavior record is necessary (see Bakeman and Gottman, 1997).

B. Conduct the study using at least four subjects in an appropriate social grouping. It is suggested that substantially more time be invested in this project.

C. Employing correlation, regression, or sequential statistical procedures (access to a computer system with the necessary software is a requirement), analyze the data set for relationships between behavior and the independent variable(s), or for sequential probabilities.

D. Present a scientific report with the appropriate tables and graphs to show support for or rejection of the hypothesized relationships.

Check Sheet For FTS Sampling

Observer:______________________________Date:________________________

Subject Group:_________________________# in group:____________________

Start Time:___________Subject Individual:______________________________

Time	Behavior	Time	Behavior	Time	Behavior	Remarks
1		1		1		
2		2		2		
3		3		3		
4		4		4		
5		5		5		
6		6		6		
7		7		7		
8		8		8		
9		9		9		
10		10		10		
11		11		11		
12		12		12		
13		13		13		
14		14		14		
15		15		15		
16		16		16		
17		17		17		
18		18		18		
19		19		19		
20		20		20		
21		21		21		
22		22		22		
23		23		23		
24		24		24		
25		25		25		

continued

Check Sheet for FTS Sampling (cont'd.)

Observer:______________________________Date:________________________

Subject Group:__________________________# in group:___________________

26		26		26		
27		27		27		
28		28		28		
29		29		29		
30		30		30		
31		31		31		
32		32		32		
33		33		33		
34		34		34		
35		35		35		
36		36		36		
37		37		37		
38		38		38		
39		39		39		
40		40		40		

Exercise 9

Focal Animal Sampling Intensive Study of Behavior

Since the methodological revolution engendered by Jeanne Altmann's 1974 paper in *Behaviour* (it is suggested that you should have read that paper before reaching this point!), the procedure known as focal animal sampling has become **the** standard technique for most primate studies.

This technique may also be called continuous sampling, but the procedure remains the same. *The essence of focal animal sampling is concentrated attention upon a single individual subject for a specific "sample period," recording EVERYTHING that the subject does or has done to it, usually with careful attention paid to the times at which behaviors change.*

The technique is amenable to the collection of both event and state behaviors. One or the other, or most frequently both, may be recorded. In this particular exercise, it is suggested that both event and state behaviors be recorded. The student must, however, be careful to separate the two kinds of data before beginning the analysis. This separation is critical, otherwise the analysis will become invalid and meaningless.

A typical focal animal sampling procedure may be limited to the recording of a specified list of behaviors that are related to a particular research design. But more commonly, the observer will record the totality of behavior both initiated by the subject and received by that individual. *The KEY advantage of focal animal sampling over all other techniques is that it allows the observer to collect the maximum amount of information on the behavior of each subject, and it is the only way to collect data on behavioral sequences without missing anything.* If sequential data are not important in the research design of the study, a simple check sheet can be constructed on which to collect the data, the development of which is left as an exercise for the reader.

Because data is collected on only a single subject at a time, the sampling process must be controlled. This can be accomplished either via a fixed schedule for a short-term study or by randomizing the observation of the subjects in a long-term study (see chapter 2 for a discussion of these aspects). One or the other mechanism *must* be employed in order to ensure that each individual in a subject population is sampled equally.

Important Factors in Focal Sampling Procedures

There are a number of important considerations involved in conducting a focal animal sampling study. Most significant for the beginning researcher are the following five items.

Ethogram: *It is important to have an appropriate ethogram available to ensure consistency in the recording of behaviors*, however, both events and states need to be defined. At this point the student researcher may wish to retire to the library to examine the limited number of published ethograms available. If a species-specific ethogram is unavailable, the most appropriate starting point is an ethogram from the most closely related species available.

Exhaustive categories: It is most desirable to have exhaustive and mutually exclusive behavior categories. "Exhaustive" means that the subject is always recorded as being involved in something in order that a complete time record can be produced.

Mutually exclusive: This means that there is NO overlap in the behavioral categories such that the subject is recorded as doing two or more activities simultaneously. One of the problems with this latter injunction is that it becomes necessary to generate a number of "combined" behaviors in order to record "walking and chewing gum" type of activities. A sometimes more useful solution is to allow the recording of simultaneous activities, but this requires extreme care in the analysis of them.

Recording codes: A compressed scoring code is of great advantage. It is far faster and simpler to record "GGam4" rather than writing out "gives groom to adult male number four." It is also important to have codes for "out of sight," and "indeterminable activity" in order to achieve an exhaustive record. It is important to record the codes with the behavior definitions in a codebook.

Active and passive behaviors: One very important factor that has been a stumbling block for beginning observers in the past is the problem of how to deal with the situation in which the focal subject is the recipient of some behavior, not the active but the passive performer in the exchange. The solution is obvious to some workers, but not others. The obvious solution is to create a set of categories in the ethogram that are "passive" or "receiver mode" behaviors, such as "receives grooming from XXX," or "is aggressed against by YYY," and any others that may be needed. In most cases behaviors that are interactive can be easily recorded as either active or passive.

The Exercise

The purpose of this exercise is the development of observational skills through the conduct of an in-depth study of the behavior of a single species group using the technique of focal animal sampling.

The exercise can be conducted according to the following steps:

A. Select a social group with a minimum of four individuals. If the student has been conducting studies on a particular species in previous exercises, now might be the time to broaden the experience envelope. Select a new species!

B. Construct an observing schedule so that each subject is the focus of the same number of samples, and as the study should be conducted over more than one day, rotate the sequence so that animals studied late on the first day are looked at earlier on the next. But be careful not to allocate the same subjects to the midday periods.

C. Conduct the sampling by means of ten-, fifteen-, or twenty-minute focal samples in which a single individual is the subject. *Everything that it does should be recorded, and the actions of all other individuals not interacting with the focal MUST be ignored.* Remember to take a short break between each sample in order to avoid mental and physical fatigue. Records may be taken as diary notes on paper, or if care is taken to be concise, then a tape recorder may be used. Be aware that use of a tape recorder is an invitation to "motor-mouth" and the development of long, rambling discourses irrelevant to the study may occur—this leads to long transcription times and can degrade the data quality.

D. Once five or ten hours of observation are accumulated, the process of analysis may be started as a check on the validity of hypotheses, design, and variable reliability. The continuance of the study can then go forward with confidence.

E. The analysis of the data must, at a minimum, contain separate tabulations of the frequencies and percentages of event behaviors and of the durations and percentages of the state behaviors observed. You are encouraged to make use of as many of the calculations in chapter 4 as are appropriate to the data and to the purposes of the study. These should be tabulated for the group as a whole as well as for the relevant individual age and sex classes.

F. The final report is to be prepared as a written scientific report in the format generally used in *Primates* and *Folia Primatologica* (see the discussion of the format of a scientific report in chapter 5, and the example found in the appendices). The report should be *no longer than twelve to fifteen typed or computer-printed pages*, including tables, graphic presentations, and appendices (obviously, a carefully handwritten report would be longer). The report text should be double spaced to allow the reviewer some space for comments and critique. The document is expected to reflect not only your direct research results, but the material acquired from reading library sources. Conclusions should also adequately support or reject similar hypotheses found in the appropriate literature.

G. An ethogram incorporating the compressed code symbols utilized and an example of a data record should be attached to the report as an appendix.

An Advanced Version of this exercise would differ only in the expectation that twenty to thirty hours of observation should be conducted. Additionally, such an advanced version may be expected to be a maximum of twenty pages in length, to reflect the results of a substantial literature search, and to aspire to the levels of a "publishable report."

Exercise 10

All Occurrences Sampling

Collecting Specified Forms of Behavior

This technique has several restrictive conditions (see the list below), but when these conditions are met, it can be an effective technique to recover frequencies and synchronization data on event behaviors. *The essence of all occurrences sampling is the direction of attention toward the entire set of subjects for a specific "sample period," recording all instances of a limited set of* highly visible *behaviors.*

The technique is amenable to the collection of event behaviors such as sexual mounts, aggressive episodes, chases, or similar behavior.

The development of an all occurrences record is deemed to be possible by Altmann (1974) if the following conditions are present and adhered to:

- Excellent observing conditions exist. That is, all subject animals must be visible and remain so throughout the sample period.
- The behaviors are highly visible. That is, the behavior itself must be so obvious that the observer will *always* see it, and hence all cases will be recorded.
- The repetition rate for the behavior is slow enough that "all occurrences" can actually be perceived and recorded.

The duration of the sampling period is not critical, but some time control is necessary in order to convert the raw frequencies of occurrences into rates of behavior. Anything from one hour to a day have been employed successfully.

Important Factors in All Occurrences Sampling

There are some important considerations involved in the conduct of an all occurrences sampling study. Most significant for the beginning researcher are the following items:

Ethogram: *It is important to have an appropriate ethogram available to ensure consistency in the recording of behaviors*; however, only highly visible events need to be defined.

Mutually exclusive: This means that there is NO overlap in the behavioral categories such that the subjects are recorded as doing two or more activities simultaneously.

Recording codes: A compressed scoring code is of great advantage. It is important to record the codes with the behavior definitions in a codebook.

Ensured observing conditions: It is critical to this procedure that all of the subjects be visible throughout the observation period. Thus this technique may require the establishment of connections with the keepers and arrangement for the group to be restricted to a single unit of their enclosure. Often this will be unnecessary in the winter months when the group may have access to only an indoor enclosure.

The Exercise

The purpose of this exercise is the development of observational skills through the conduct of a study of the behavior of a single species group employing the technique of all occurrences sampling.

The exercise can be conducted according to the following steps:

A. Select a social group with a minimum of four individuals. Now may be the time to broaden the experience envelope by selecting a new species.

B. Check and verify that your ethogram subset, the highly visible dramatic behaviors such as aggressive episodes, vocalizations, mounting, and so forth, are clearly defined and appropriate to the study.

C. Select times, or arrange with the keepers for times, when the group is restricted to a single enclosure.

D. Conduct the sampling by means of one-hour-long sample periods. Records may be taken as diary notes on paper; on a check sheet constructed for the specific project; or, if care is taken to be concise, a tape recorder may be used. But take note of the cautions expressed in exercise 9.

E. Once five hours of observation are accumulated, the process of analysis may be undertaken. Instructors may specify longer or shorter periods and overall study time.

F. The analysis of the data must, at a minimum, present tabulations of the frequencies and rates of the defined event behaviors observed.

G. The final report is to be prepared as a written scientific report in the format generally used in *Primates* and *Folia Primatologica* (see the discussion of the format of a scientific report in chapter 5, and the example found in the appendices). The document is expected to reflect not only your direct research results, but the material acquired from reading library sources. Conclusions should also adequately support or reject similar hypotheses found in the appropriate literature.

Exercise 11

Matched Control Sampling
Studying Reconciliation Behavior

This technique has been developed over recent years as a method devoted to the study of reconciliation behavior, a topic first brought to prominence by de Waal and van Roosmalen in 1979. Reconciliation was a hot topic in primatology during the late 1980s and 1990s and is the focus of a large body of new literature. The concept is based upon the premise that for a social group of primates to maintain its existence, there must be mechanisms to reestablish friendly relations after a conflict situation has developed. This implies that some behaviors or interactions serve to reduce anxiety to baseline levels, and these can be categorized as affinitive behaviors showing social attraction between the participants.

> Affinitive contact between former opponents soon after a conflict has been demonstrated in a growing number of primate species. Several recent studies show that such contact reduces the probability of future conflicts, allows the recipient of aggression to reduce its anxiety, and restores tolerance between former opponents. Hence, these contacts can be termed reconciliation. (Kappeler and van Schaik, 1992:51)

However, reconciliation has been operationally defined using different criteria in different species by different observers, and no single definition is universally recognized. Functional reconciliation, on the other hand, can be clearly defined as "behavior that restores a dyadic social relationship" (Cords, 1993:255). Cords goes on to show that for long-tailed macaques (*Macaca fascicularis*), the operational reconciliation can be defined as:

> (i) first post conflict non-aggressive encounters between former opponents, including mere proximity, (ii) occurring after a conflict sooner than expectations based on baseline interaction rate measured (once) for the same dyad at about the same time, (iii) regardless of which opponent initiates the encounter. (Cords, 1993:264)

Cords sees the above as agreeing closely with the functional definition of reconciliation. It must generally be conceded that there will be differing behaviors employed as reconciliation in different species, and not all species display more than a minimal frequency of the phenomenon. Thus, in most studies of reconciliation, the first task has normally been to establish that it exists. It is only then that the observer can proceed to study it.

The technique of matched control sampling is distinctive in that it incorporates a control sample against which is tested another sample. In reconciliation studies, this test sample is a postconflict period, triggered by the observation of an aggressive conflict (but in theory any behavior could be examined with this method). Hence the frequently encountered reference in the literature to "PC/MC," or postconflict, matched control study paradigm.

The pattern for MC sampling is as follows: A trigger behavior is established as the signal to begin a sample observation. In reconciliation studies this is usually an aggressive conflict between two individuals. At this point the observer begins a ten- or twenty-minute focal animal sample on *one* of the combatants. This is random, perhaps decided at the moment by the flip of a coin (e.g., heads—subject A is focal, tails—subject B is focal), or by a preestablished protocol (e.g., A has more samples recorded than B, select B as focal next time there is a conflict between A and B), or by an observing rule (e.g., always observe the aggressor in a conflict).

The second component—the matched control—is a second focal animal sample conducted on the same subject but twenty-four hours later, starting at approximately the same time as the postconflict sample began. Many researchers have added specific criteria regulating the start of the MC sample. Some have employed a spatial criterion, requiring the two subjects to be within a specific distance of each other, or in the same subgrouping. These criteria may have greater validity in free-ranging animals than in caged groups (Kappeler and van Schaik, 1992). However, the most common criteria have been limited to the timing and the absence of agonism between the subject pair in several minutes before the sample start. The method of observation is simple as regular focal animal sample records are collected. The difficulties come with the analysis of the data, and two techniques have developed: one in which there is a specific comparison made between the PC sample and the corresponding MC sample, and a second in which a base-line level of interaction and proximity is based on aggregate MC samples (Kappeler and van Schaik, 1992). Silk (1997) has further criticized the MC methodology by pointing out that, "If conflict damages social relationships and these effects persist over time, . . . then it seems inappropriate to conduct the matched control observation on the day after conflict has occurred." Also, "A better alternative would be to collect baseline data that can be retrospectively matched with post-conflict observations." (p. 267).

As can be seen, there are a number of cautions that adhere to the matched control sampling paradigm. Care and caution must be exercised in any matched control study. The exercise will focus upon the main variants of MC analyses and not require the construction of a baseline data collection.

Analysis of MC Sampling

The following discussion of and diagrams depicting analytical approaches to PC/MC data are based upon Figure 2 and its caption in Kappeler and van Schaik (1992).

Diagram A presents the basic comparative framework for PC/MC studies, and Diagram B shows the first analytical procedure. The *conciliatory tendency method* relies only upon the PC data and presents results as "a reconciliation" (represented by the black dot in Diagram B) if affinitive behavior occurs at all. The only quantitative analysis possible is to measure the time between conflict and reconciliation, and then present an average for group or age-sex classes. Decisions about whether or not the conflict is resolved in this method are very dependent upon the length of the sampling period.

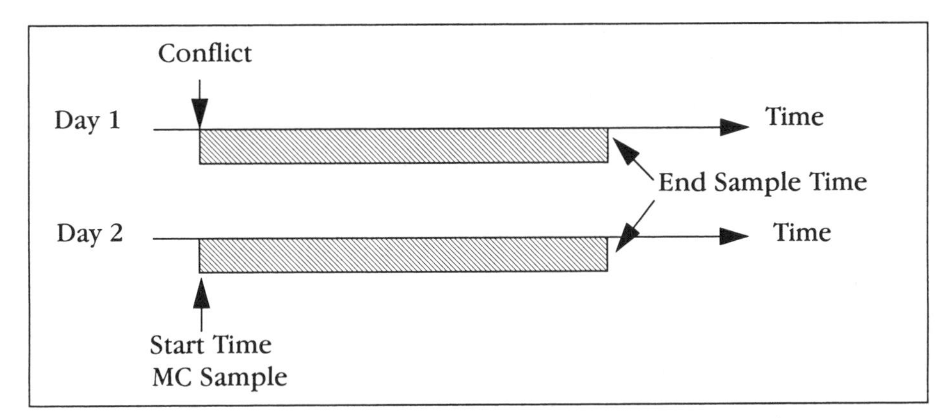

Diagram A Basic comparative framework for PC/MC studies.

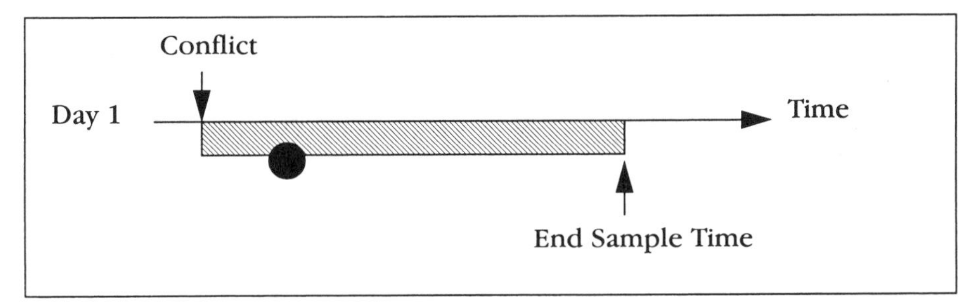

Diagram B Conciliatory tendency method.

Diagram C presents the *conservative reconciliation method*, which only considers a conflict to be resolved if affinitive behavior occurs in the PC sample but *not* in the MC sample. This technique can produce false negative evaluations if affinitive behavior actually occurs in the MC sample (Kappeler and van Schaik, 1992).

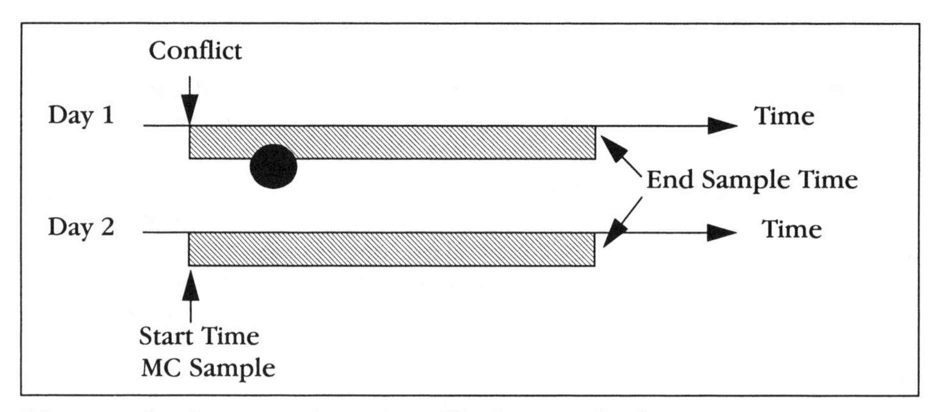

Diagram C Conservative reconciliation method.

In Diagram D, the *attracted pairs method* is diagrammed. All conflicts that are followed by an affinitive interaction are categorized as "reconciled" if the affinitive behavior occurs earlier in the PC sample than in the MC sample, and as "unreconciled" if later than in the MC. This technique also has some limitations derived from probability assumptions, and produces the occasional false positive and false negative evaluations.

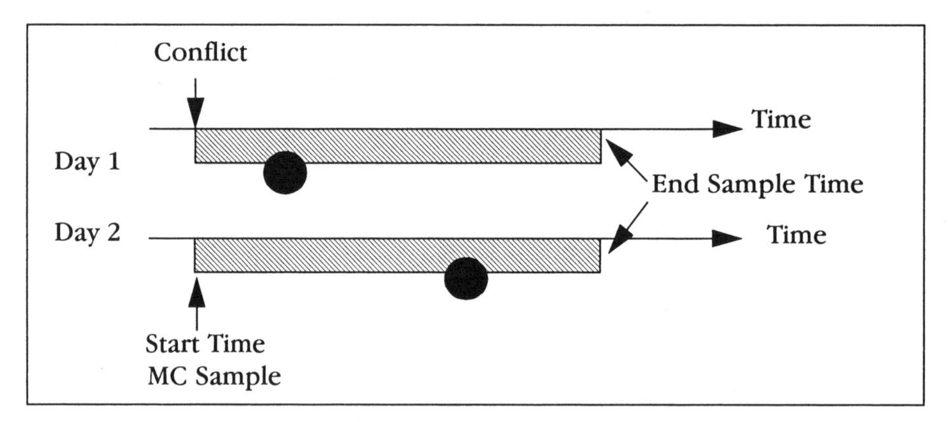

Diagram D Attracted pairs method.

The *n-minute* or *time rule method* shown in Diagram E applies a rule derived from a larger set of control samples. Basically a large number of sample periods are set up as control periods and the rates at which affinitive interactions occur are used to decide the time rule. This is beyond the requirements of the exercise, and for practical application, I suggest that a "two-minute rule" be used for this analysis. This means that if the affinitive behavior occurs earlier than two minutes after the conflict, the conflict is categorized as "reconciled," if later in the sample, it results in a categorization of "unreconciled."

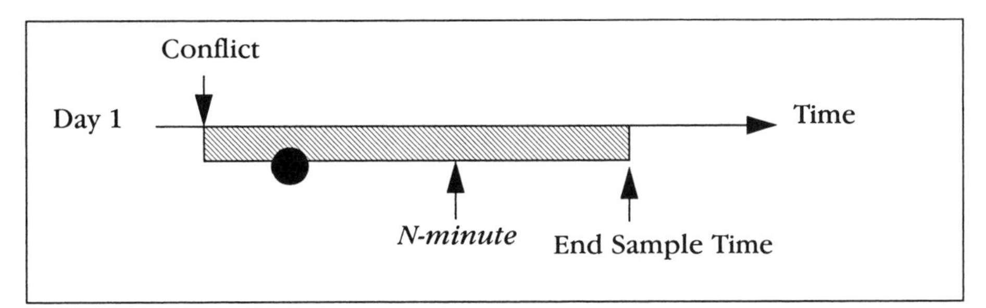

Diagram E N-minute or time rule method.

The final component in this analysis chain is then a calculation of the percentage or proportion of reconciled versus unreconciled conflicts. As always this may be broken down by age-sex class or even to individual levels.

THE EXERCISE

The purpose of this exercise is to conduct a study of the behavior of a single species group employing the technique of postconflict, matched control sampling.

The exercise can be conducted according to the following steps:

A. Select a social group with a minimum of four individuals. The larger the group is, the more opportunities exist for the study to obtain adequate data.

B. Check and verify that your ethogram is clearly defined and appropriate to the study. An important preliminary is to establish that the species actually has affinitive behaviors that serve as reconciliation behavior, and that these occur frequently enough to be useful.

C. Conduct the sampling by means of watching the study group until a conflict takes place—immediately note the time and begin a focal animal sample on one of the subjects involved (this individual must be clearly identified). Records may be taken as diary notes on paper, or if care is taken to be concise, then a tape recorder may be used, but take note of the cautions expressed in exercise 9.

D. On the following day continue with informal watching of the study group, but pay careful attention to the individuals involved in the interaction of the previous day during the fifteen minutes prior to the time designated for the start of the MC sample; if conflict does not recur, the MC sample may proceed to be recorded. However, if a conflict takes place during this period, the MC should be postponed until the following day. (Note that this may result in a chain of postponements and is one of the factors that Silk [1997] criticizes.)

E. The analysis of the data must present tabulations of the number of conflicts and reconciliations observed. It should also make use of the conciliatory tendency, conservative reconciliation, attracted pairs, and *n-minute* rules to compare the different analysis methods.

F. The final report is to be prepared as a written scientific report in the format generally used in *Primates* and *Folia Primatologica* (see the discussion of the format of a scientific report in chapter 5, and the example found in the appendix). The document is expected to reflect not only your direct research results, but the material acquired from reading library sources. Conclusions should also adequately support or reject similar hypotheses found in the appropriate literature.

Testing and Analysis
Tools for Understanding Behavior

Exercise 12
Inter-Observer Reliability Testing:
Do Two People See the Same Things?

Exercise 13
Analytical Structures:
Interaction Matrices and Sociograms

Exercise 12

Inter-Observer Reliability Testing
Do Two People See the Same Things?

Any science that depends upon human perception as its main source of data has a very large problem to resolve. The deficiencies of human perception place in question the reliability of all research in primatology (and by extension, all other observational sciences). Fortunately, awareness of these deficiencies is, in itself, half of the solution to the problem. If one is aware of the problem, then steps to either correct or make allowances for it can be taken and objective evaluations can be made.

The correction of perceptual biases, and so forth, is an internal process unique to each observer; how then can anyone place any degree of confidence in the results that someone else has produced? The solution lies in the realm of reliability testing.

Reliability Testing or Observer Agreement?

Bakeman and Gottman (1997) point out that there is some disagreement about the meanings of "observer agreement" versus "observer reliability." Citing Johnson and Bolstad (1973), they argue that agreement is a general term describing the extent to which two observers agree with each other. Reliability is more restrictive and derived from psychometrics where it is used to demonstrate how accurate a measure it is. Reliability for an observer can only be assessed against a standard protocol.

Reliability of results can be dramatically improved through the adoption of appropriate methods of data collection. Achieving a high level of competence in observation is the object of these exercises. Control of methodology and utilization of realistic research design can significantly improve the quality of data that an individual observer collects, but we must remain aware that all observations are still filtered through a human brain—and one human brain that may or may not see things in exactly the same way as other human brains. The solution to this dilemma has been available for quite some time in what is referred to as "interobserver agreement testing." A similar process is also used to ensure that single observers do not drift in their perceptions and definitions of behaviors over the course of their project—more about that later. It ought to be apparent that the simplest way to test one's observational reliability is to make a concurrent check to find out if what one observer records as "behavior *A*" is the same as what another observer records. Over the years this has evolved into a regularly used coefficient of reliability *R*, a proportional value that equals 1.00 when the two observers are in perfect agreement.

$$R = \frac{s}{t/2}$$

In the equation, R is the coefficient of reliability, s is the number of same behavior records scored by two observers, and t is the total number of behaviors collected by the observer pair.

It is generally considered that a value of $R = .98$ or better is needed to consider the results of two observers to be comparable in any useful way. R is tested and determined in the following manner:

A. Two observers who are to check their reliability set up to make a "test observation for reliability." This is done by the two workers independently observing the SAME SUBJECT at the SAME TIME for the SAME OBSERVATIONAL PERIOD.

B. Typically the two workers will independently record continuous data, that is, via focal animal sampling, for a fixed period (usually ten or fifteen minutes is sufficient) on the same subject. Alternatively, or to facilitate a quick test, focal time sampling is an appropriate procedure to follow, especially if only state behaviors need to be validated.

A critical condition for reliability testing is that *during the test there must be* NO *communication between the observers, although they should be relatively close together in order to have the same angles of view.*

C. At the end of the test, the two compare their records, side by side, and calculate their coefficient of reliability—R.

D. R can be calculated by adding up the number of "same behavior" recordings (s in the formula above), dividing by half the total number of behaviors recorded by both observers ($t/2$ in the formula). It is often then multiplied by 100 to obtain a percentage agreement value instead of a proportion. An alternative formulation that may be easier to calculate uses the number of agreements—N_a—and the number of "disagreements"—N_d—in the formula:

$$R = \frac{N_a}{N_a + N_d}$$

and again it may be multiplied by 100 to obtain a percentage agreement.

E. As an example, where two observers recorded 139 total behaviors, Observer One with 70 records and Observer Two with 69 records of which 66 are the same as Observer One's, the calculation is as follows:

$$s = 66 \quad t = 139$$

$$R = \frac{s}{t/2}$$

$$R = \frac{66}{139/2} = \frac{66}{69.5} = .9496$$

and the percentage agreement can be rounded off to 95 percent.

In the alternate calculation, N_a is 66, and N_d is 4 (three direct coding differences and one behavior missed by Observer Two), the value calculates to .9428 for a percentage of agreement of 94 percent. It is important to note that the two formulae will give slightly different results.

$$R = \frac{66}{66 + 4} = \frac{66}{70} = .9428$$

F. Since only 94–95 percent agreement is not very good for professional observers (but it is better than average for beginners), the two would proceed to compare their behavioral categorizations and discuss what it was that they saw differently. The reasons for the discrepancies may be immediately apparent to the two workers, or it may take some time to find them. In this example, since the counts of behaviors seen by each observer differ (70 versus 69) by one, it is possible that Observer Two was inattentive and simply missed an item. On the other hand, the situation might be more complicated, with Observer One making finer distinctions within a certain behavior set that Observer Two considers to be all the same thing, thus yielding the three items different between their records. In any case the test is repeated and repeatedly discussed until the two are satisfied with their R score.

In many team-based research projects such reliability testing is built into the schedule so that each observer of a team is tested against every other as a matter of course each week or every other week. This serves the important function of maintaining behavioral category agreement among all the members of the observing team over a sometimes lengthy study period, assuring that the behavioral categories remain the same from beginning to end, and from observer to observer. Such procedures fall into the category of true reliability testing and have the added advantage that old observers can leave and new ones can be incorporated into the study team without the loss of data or the distortion of behavioral categories.

For the individual observer conducting a long-term project, the matter of "reliability testing" is not one that can be sloughed off and ignored. Instead, it is one of those activities that can prove to be vitally important to the validity of the entire data set. The problem for such an individual researcher is that there is no one to test against, and, hence, the person must resort to a form of self-testing, one that has a significant commonality with the pretest/posttest paradigm used in psychology and medicine. Under this self-testing procedure, the observer conducts a sample test at the beginning of the study (but after the initial test and revisions of design have been completed) and prepares a formal ethogram description of the behaviors. Periodically thereafter, the observer takes sample tests for comparison to the original record, and the ethogram. It can be most effective if the level of description employed in the testing is at the atomic level. Periodic examination of the ethogram may help to maintain consistency. Another alternative, where video equipment is available, is for the observer to prepare a ten- or twenty-minute live record, and then to periodically play the tape and record data from it as if seeing it for the first time. The observer can then compare new with past records and calculate the R between the samples. Utilization of one or the other of these variations is an aid to maintaining consistency in the quality of data.

The Exercise (Simple Version)

The purpose of this exercise is to conduct an evaluation of the interobserver agreement between two observers.

A. Two observers will set up an ethogram or preferably use an established ethogram set of behaviors and agree to conduct an observational test on an appropriately active species, using either focal time sampling (ten-minute sample period with ten-second intervals = sixty observations is suggested) or focal animal sampling for a ten-minute period.

B. The "test" should be conducted twice, with an evaluation stage in between each test to correct perceptual errors. This will be relatively easy if focal time sampling is being used, and consequently the records may be matched observation by observation to locate disagreements. The process will be slightly more complex if using focal animal sampling since event behaviors may also be recorded but the derivation of a behavioral sequence is straightforward.

C. The two observers jointly discuss the behavior sequences recorded and note the omissions and perceptual variations in the data. Note that if comparing lists on a side-by-side basis, and one observer has missed a behavior, then all of the sequence will be misaligned from that point. The omission should be identified and the alignment corrected before proceeding to calculate the *R* value.

D. The coefficient *R* is to be calculated after each stage and recorded. If the two observers are in perfect accord after the first test (i.e., $R = 1.0$), the two observers may choose to conduct the second test as focal animal sampling (continuous sampling) over a twenty-minute sample to extend their self-evaluation.

E. Finally, a report may be prepared in the form required by the instructor.

The Exercise (Advanced Version) and Cohen's *Kappa* Calculation

Bakeman and Gottman (1997) make a number of severe criticisms of the use of agreement percentages as an estimate of observer reliability. They point out that there is really no rational basis for considering a percentage agreement value in the 90+ percent range to be good. They also note that the number of codes used can influence the percentage agreement, and, consequently, comparability is lost between studies. The two authors go on to provide a most telling argument against its use:

> Given a particular coding scheme and a particular recording strategy, some agreement would occur just by chance alone, even with blindfolded observers, and agreement percentage scores do not correct for this. (Bakeman and Gottman, 1997:61)

What is the solution to these defects? Bakeman and Gottman devote most of chapter 4 to this question with a thorough examination of the utility and validity of Cohen's *kappa* as applied to observer agreement. The *kappa* statistic corrects for chance agreement but requires the generation of an "agreement" or "confusion" table that shows graphically where similar and dissimilar decisions by two observers occur.

Figure 12.1 demonstrates the cases in which agreement and disagreement between two observers took place in a test example. Disagreements occurred in each behavior category but a percentage agreement value (*R*) of 87 percent can still be obtained.

With this data in hand, it is possible to calculate the value of *kappa*. We will deal with only the basic calculation; if correction factors are needed, the reader is referred to the detailed discussion in chapter 4 of Bakeman and Gottman. The basic *kappa* formula is:

$$K = \frac{P_{obs} - P_{exp}}{1 - P_{exp}}$$

Two additional calculations are needed to derive the value of *kappa*, and they can be calculated from the confusion table values. P_{obs} is the observed proportion

		Second Observer					
		Unoccupied	Solitary	Together	Parallel	Group	Totals
First Observer	Unoccupied	7	0	1	0	0	8
	Solitary	1	24	0	0	0	25
	Together	1	1	17	2	2	23
	Parallel	0	0	3	25	1	29
	Group	0	0	0	1	14	15
	Totals	9	25	21	28	17	100

Figure12.1 Agreement or confusion matrix table from Bakeman and Gottman (1997:62). Values on the diagonal indicate agreement and off-diagonal values show the number of cases of disagreement for each of the five categories of behavior used in the test study.

of agreement and is calculated by summing the values in the diagonal and dividing by the total number of observations. This is (7 + 24 + 17 + 25 + 14)/100, so P_{obs} = .87. Similarly, the value of P_{exp} is the proportion that is expected by chance. It is calculated by summing the products of the marginals. The total for row 1 is multiplied by the total for column 1, added to the product of row 2 by column 2, added to the product of row 3 by column 3, and so forth, until the entire matrix is used, then this value is divided by the square of the number of observations.

$$\text{Thus } P_{exp} = \frac{(9 \text{ x } 8) + (25 \text{ x } 25) + (21 \text{ x } 23) + (28 \text{ x } 29) + (17 \text{ x } 15)}{100 \text{ x } 100} = .2247$$

$$\text{And therefore } K = \frac{.87 - .2247}{1 - .2247} = .8323 \text{ (after rounding)}$$

One of the advantages of *kappa* is that the distribution is known and a *kappa* value can be tested for significance. This is beyond the needs or expectations of this workbook, and the reader is again referred to chapter 4 of Bakeman and Gottman (1997).

The advanced exercise is exactly the same as the simpler version except that the calculation of *kappa* is required. The form on the following page (also on the CD) may be used to work through these exercises. An excel worksheet for calculating Cohen's kappa can be fond on the CD in the worksheet folder.

Reliability Testing: 1

Name:________________________ Partner Name: ________________________

I.D.#:________________________ I.D.#:________________________

Species/Group Observed: ____________________________________

Test Date:____________________ Time:____________________________

REPORTER'S CODES	PARTNER'S CODES
CALCULATION OF ***R or kappa***	PERCENTAGE AGREEMENT

Reliability Testing: 2

Name:________________________ Partner Name: ________________________

I.D.#:________________________ I.D.#:________________________

Species/Group Observed: ____________________________________

Test Date:____________________ Time:________________________

REPORTER'S CODES	PARTNER'S CODES
CALCULATION OF *R or kappa*	**PERCENTAGE AGREEMENT**

Exercise 13

Analytical Structures
Interaction Matrices and Sociograms

The collection of data is only half of the research paradigm for observational primatology. The second component is the analytic process. For many researchers this may consist of transcribing the raw data into a computer and making use of any of a number of standard statistical packages such as SPSS, SAS, SYSTAT, MINITAB, and so forth, to produce tables and graphs of results. These statistical packages are available both on mainframe computers and as versions to run on Windows and Macintosh computers. There are also various packages such as CricketGraph, Excel, Lotus 1-2-3, Paradox, Advantage, and so forth, that can also perform analyses.

However, there are two older analytic methods that are both simple to employ and effective in providing insights into the organization of a primate society. These are interaction matrices and sociograms.

Interaction matrices and sociograms both serve as visual tools of analysis and as presentation devices. By this we mean that an interaction matrix or a sociogram serves as a tool during the analysis phase and can also be employed as the primary means of presenting the results of matrix analysis.

Interaction Matrices

An interaction matrix, sometimes called a sociometric matrix, is simply a two-axis table that is organized with the actors along one axis and the recipients along the other. Note that each matrix is restricted to one specific behavior pattern. Thus to begin with a matrix analysis it is necessary to have data for both the actor who emits the behavior and the recipient, who is the receiver of the behavior. This means that it is also necessary to have, at minimum, category-level identification, and preferably individual recognition, in order to be able to make records in the form "George (male 321) aggressed against Ralph (male 456), who fled." This is the kind of data that is collected in focal animal or continuous sampling (exercise 9). If this type of data is available, then it is relatively simple to construct the matrix. An empty matrix for a particular behavior pattern (we will use grooming among males for this example) would appear as in Figure 13.1.

Recipients

	X	Joe	Fred	Alf	Moe	Fuzz
Actors	Joe	X				
	Fred		X			
	Alf			X		
	Moe				X	
	Fuzz					X

Figure 13.1 Example of an empty matrix.

We would then proceed to identify the individuals across the top (the column labels) as the groom*ees*, and the same sequence of individuals down the left column (the row labels) as the groom*ers*. Since it would be difficult to evaluate grooming between the same individual as groomer or groomee, the diagonal is filled with a null indicator for that cell. The next stage is the critical component of the analysis. One goes through the data expressly searching for grooming episodes, each must involve known individuals and have an indication of which role each individual was engaging in, and a tick mark is placed in the appropriate cell. An alternative data collection technique, all occurrences sampling (exercise 10), could be used if interaction matrices are to be the only analytic tool employed, but it is probably more useful to perform an after-the-fact extraction of occurrences from focal animal sampled data. In this pattern of data examination, if Moe, being the focal subject at the time, is counted as being a groomer in an episode with Fred (naturally the groomee), a tick mark goes into the box in the first column where Moe's row intersects Fred's column. If Alf is the focal and he grooms Joe, a tick goes under Alf's row in Joe's column, and so forth. After a while the matrix may resemble the one in Figure 13.2:

Groomee

Groomer	X	Joe	Fred	Alf	Moe	Fuzz
	Joe	X	✓✓		✓✓	✓
	Fred		X			
	Alf	✓✓✓	✓✓✓✓	X	✓✓	✓✓
	Moe		✓✓		X	
	Fuzz					X

Figure 13.2 Example of a matrix using tick marks to show grooming episodes.

Eventually the tick marks will be converted to numbers at some point, and a final matrix might look like the following:

Groomee

Groomer	X	Joe	Fred	Alf	Moe	Fuzz
	Joe	X	21		29	11
	Fred		X			
	Alf	22	11	X	8	18
	Moe		11		X	6
	Fuzz		2			X

Figure 13.3 Example of a matrix using numbers to show grooming episodes.

This, however, would make a very cluttered presentation, and it is difficult to decipher quickly. In order to make the presentation more readable and to increase the impact, the cells of the matrix should be shuffled up and down by rows, then left and right by columns, such that the left side reflects a sequence of increasing or decreasing frequency as groomer. The net result of this activity will be to place all or most of the zero cells in one triangle or the other and concentrate the filled cells on the other side. For presentation the matrix in Figure 13.3 would become the one in Figure 13.4. It ought to be evident that Fred receives the most grooming and does the least.

Groomee

Groomer	X	Alf	Joe	Moe	Fuzz	Fred
	Alf	X	22	8	18	11
	Joe	0	X	29	11	21
	Moe	0	0	X	6	11
	Fuzz	0	0	0	X	2
	Fred	0	0	0	0	X

Figure 13.4 Example of a matrix with the data arranged for presentation.

This is an idealized situation; in many cases it is impossible to obtain such a clear lower half (or upper half) to the matrix. These situations arise when there is a reversal or uncertainty of relationships, or the interaction pattern is not as completely correlated with a dominance structure as it is in this case. Here we have a clear indication that Alf does the most grooming, receiving the least, while Fred does the least, and receives the most. Under a standard dominance hierarchical interpretation, it would be suggested that Fred is the most dominant or "alpha" male, while Alf is most subordinate.

An additional advantage of an interaction matrix is that it is very simple to apply inferential statistical evaluation to one as it generally fulfils the expectations for a chi-square analysis.

SOCIOGRAMS

A sociogram is essentially a visual graphic method of analyzing the same kinds of data that can be placed into a matrix. In this case the mechanics are slightly different but the principles are the same. The procedure is straightforward; circles are drawn to represent the individual animals of a small group or of a subgroup within a larger unit. Then the records are searched for interactions of specific type, between the individuals represented. This, again, means that a sociogram is limited to the depiction of a particular subset of the behaviors (usually a single type of behavior per sociogram) observed within the study group. The interactions are represented by a line from the initiator of the action to the recipient. If there are reciprocal interactions, then these must also be represented as arrowed lines as well. Thus, a first-stage sociogram for allogrooming might look like the following illustration.

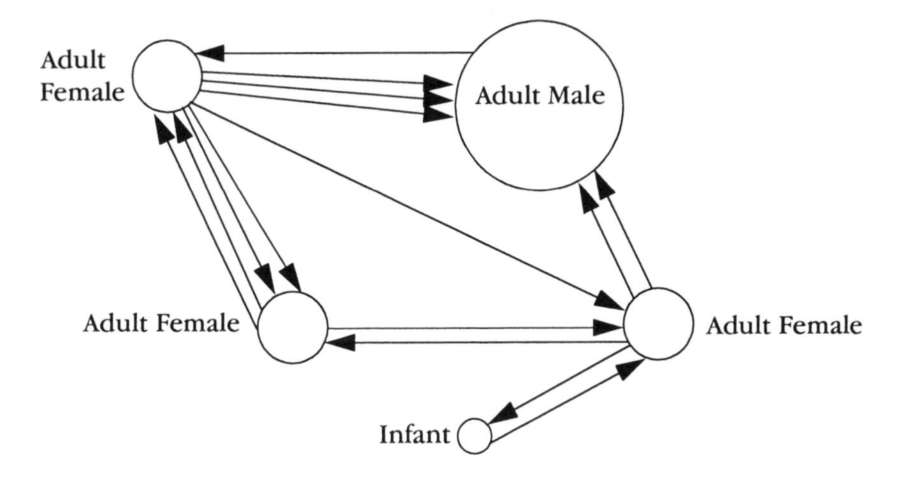

Figure 13.5 Example sociogram, each line with an arrow point represents a number of interactions.

It should be obvious that the repeated application of a line for each observed instance to indicate relationship interactions will rapidly become unreadable, and the meaningfulness of the diagram will be lost. The solution to this is straightforward, the individual lines are replaced by single heavier, proportional line representations, often including a numerical value for the frequency or duration of the interactions in each direction. Such a second-generation sociogram might look like Figure 13.6.

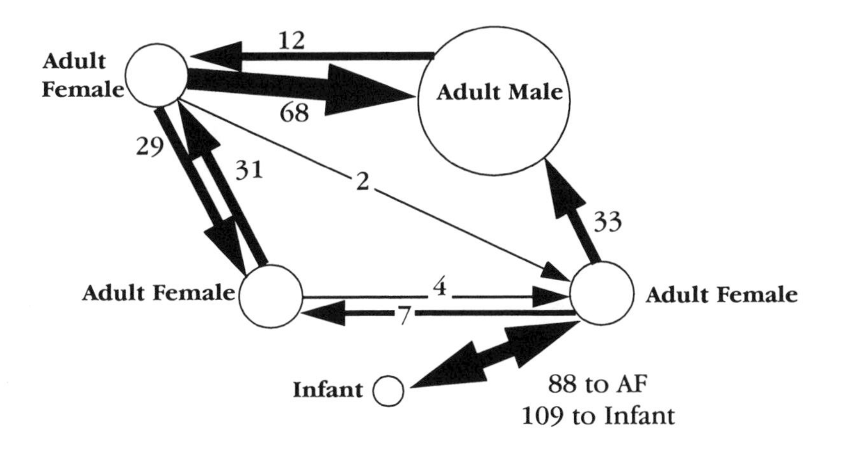

Figure 13.6 A second-generation sociogram using proportional lines and numbers to indicate the extent of interactions between the individuals.

Sociograms are also used occasionally to show probabilities of interrelationships between individuals or between specific classes of group members, but the calculation of such values is beyond the intention of this exercise. Interested students are referred to Stuart Altmann's (1968) paper in *Behaviour* for this technique.

The end result of sociogram generation is a graphic picture of the pattern of relationships amongst the members of the group depicted. This will provide information as to the direction and intensity of interaction, show which individuals are important within the unit, and show which animals are social isolates. Sociograms are valuable tools in the analysis of social organizations as well as very suited to use as presentation graphics.

THE EXERCISE

The purpose of this exercise is to develop some skill and facility in generating analytic structures. For this reason both a matrix and a sociogram are to be constructed and presented in final form. It is most appropriate for this exercise to make use of the data collected by focal animal sampling, and consequently it may be appropriate to employ this exercise in conjunction with one of the methodology training exercises. In that case, the matrices and sociograms generated would be included as part of the "presentation of results" section of the report. If this exercise is conducted separately, either on data collected by the student specifically for it, where all occurrences sampling might be appropriately used, or on a prepared data set provided to the student, the product should be a standard scientific report.

Ecology Field Exercises

The following set of projects is intended to help clarify some of the research work that is normally a component of field studies. These exercises are intended to obtain basic data about the environment and the variability of that environment as well as how primate populations interact with the common plant species of their range.

Exercise 14

The Line Transect Survey

How Many Primates Live in the Area?

Estimating how many individual primates reside in an area is one of those apparently simple problems that can become supremely confusing. Variations in degree of social aggregation, size, cryptic coloration and/or cryptic behavior, and general environmental complexity all contribute to the ease or lack with which primates are located and counted. It is clearly much easier to find and count primates living in a short-grass savanna like Nairobi Park than it is do so in the Kibale or Budongo Forests where 30 to 45 meters of tree growth is typical. It is even more difficult in a mixture of tall-grass savannas and woodlands where ground level visibility is often less than a hominid arm's reach.

While other methods exist (Anderson, Laake, Crain, and Burnham, 1979; Wilson and Wilson, 1975), the most successful and useful method of censusing primate populations has proven to be the line transect survey (Brockelman and Ali, 1987). The line transect survey method does not rely upon the existence of a defined plot, but does require some form of a trail, the straighter the better, through the environment. This trail may be a game trail, a local footpath, a road, a seismic survey line, or part of a transect grid maintained by a research facility. The researcher walks a transect and keeps a record for each primate group (and each nonprimate, perhaps) of the distance to the animals (for groups an approximation of the group center is used), the distance of the animal from the path (at right angles to it), and the angle of observation (from observer to subject relative to the path). This means that the equipment needed for a line transect survey consists of a notebook, a rangefinder, and a compass. Today, one might add a global positioning system receiver and possibly a laser rangefinder to enhance accuracy.

Initially on a line transect survey the only assumption that must be made is that all animals directly *above* the trail will be detected. This assumption may generally hold for most species, but for the cryptic forms, it will not, as they can be very difficult to locate under any circumstances. There are three other assumptions that may be violated to some degree but are generally accepted as applicable:

- Animals or groups are seen *before* they move away or flee.
- Distances and angles are measured *accurately.*
- Sightings are *independent* events.

To some extent, these assumptions may be related, particularly the first and third of these. If subjects are first seen at angles of 45° or more relative to the line of the trail ahead, the probability is that they are already in motion. If animals move, and particularly as they move away from the observer, there is a possibility that they will be recontacted, in which case, the observation is *not* an independent event. The second contact under these circumstances should not be counted into the census. Another problem surfaces if individual animals are counted rather than groups; the set of observations on the members of a group are *not* independent of each other. According to Brockelman and Ali (1987), there is no solution to violating this assumption, and it must simply be tolerated in the data set.

Observers may display a tendency to approximate, rather than measure accurately, angles and distances. This introduces artifacts into the records and must be avoided. Accurate measurement of the angle to within a degree using a lensatic compass or the more elaborate pocket transit type, and measurement of the distance to within a meter, should be performed with tape or rangefinder.

DATA REQUIREMENTS AND THE CALCULATION

As suggested earlier, the observer walking a transect must stop the moment a primate or group of primates is observed and immediately take two measurements. These are the angle between the forward direction of the transect line and the direction of the animals (usually signified by the Greek letter theta [θ], and the distance to the primate or group center (usually recorded as *S*). Thereafter other census items can be collected (species, number, type of plant/animal being consumed if feeding, height above ground, etc.). The observer then moves along to a point at which the group's position is "normal" (at 90°) to the transect track, and again measures the distance to the subjects (usually recorded as value *P*). This results in redundancy since any two of the three measurements can be used to calculate the other, but also allows the position of the primate or group to be confirmed. The common calculation is the use of the angle θ and the distance *S* to calculate *P.* Figure 14.1 shows the relationship of the values and the formula for calculation of *P.* Sine values for angles are commonly available in any book of mathematical tables and can be generated on a scientific calculator. In addition, *S* can be calculated by distance $P/\sin(\theta)$.

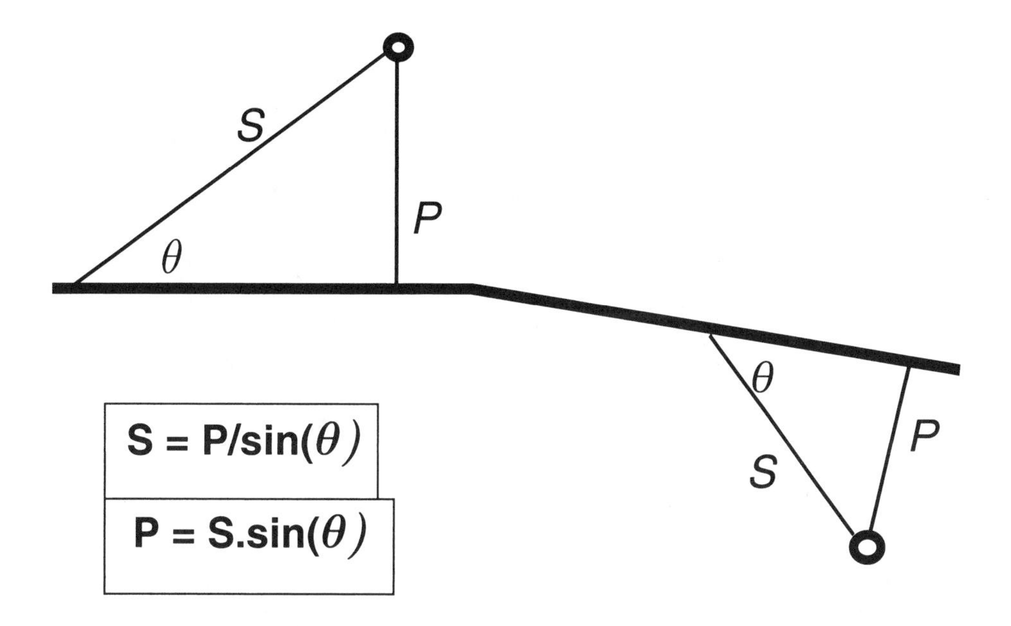

Figure 14.1 Transect line with locations of two primates, the angle θ and the distances *S* and *P.* The formula for determining *P* is distance *S* times the sine of θ, while that for obtaining *S* is *P* divided by sine of θ.

Before moving to the utilization of this data, it is necessary to touch on the design of surveys as this influences the calculation of population densities. Anderson et al. (1979) suggest that it is necessary to have some advance information on the biology of the populations being surveyed and to have an idea of the size, shape, and type of habitat being covered. They recommend that transect lines should be spaced apart so that no subject can be visible from more than one line. They recommend that the line(s) should be long enough for a minimum of forty, and preferably sixty to eighty, sightings to be collected during one cycle through the area. In areas where a research grid has been established, the lines are often parallel and spaced at 100-meter intervals—this fulfills the basic requirements, but the total lengths of transects and the encounter rate may not fulfill Anderson et al.'s minimum sighting rate. Other options include setting a suite of transect lines with random starting points and a common but randomly selected direction, or where topology of the area dictates, a series of "piecewise linear transects" might be required. These two options are diagrammed as Figure 14.2. In either case the length of the survey is obtained by adding up all of the transect lengths employed, and thus, all segment lengths need to be measured.

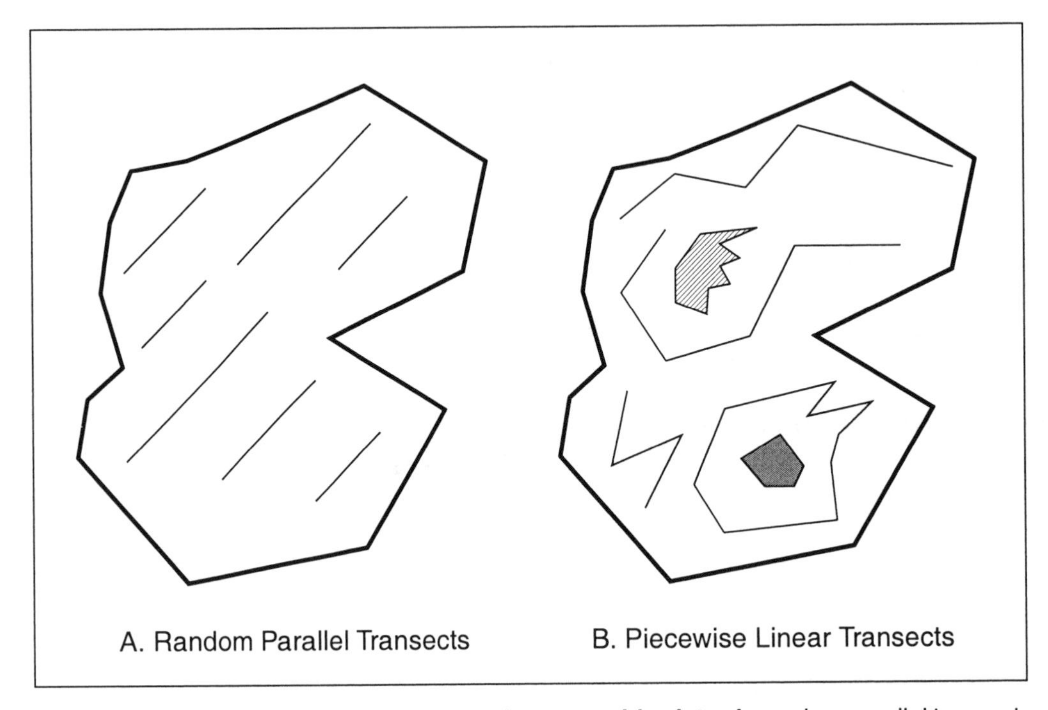

Figure 14.2 Alternate transect layouts where no grid exists. A: random parallel transect lines with a common direction—useful in a homogeneous habitat. B: piecewise transects as dictated by a rugged or nonhomogeneous environment. (Redrawn after Anderson et al., 1979:76)

Once the method of transect selection has been established, only the conduct of the survey remains, and estimation of the "width" of the transect. This estimate may be left undetermined, as recommended by Anderson et al. (1979), and essentially considered to be infinite, but in practical terms is most often dictated by the maximum distance at which it is possible to spot primates from the trail. While this value is exceedingly variable due to the range of leaf densities and vegetational complexity, the normal assumption is that it will correspond to the largest measured distance from the track. This implies that where a window through the vegetation provides a view out to significantly greater distance, these outliers should be removed from the data set.

After the data has been collected, the density (D) of the transect sampling is calculated through the formula:

$$D = \frac{N}{2Lw}$$

where N = number of groups or individuals encountered, L is the sum of the transect lengths, and w is the width on each side of the transect that is covered. Effectively this is only an *estimate* of density, and awareness of the problems inherent in this calculation should be achieved before publication of the data. The works of Brockelman and Ali (1987), Wilson and Wilson

(1975), and Anderson et al., (1979) should be consulted to understand these problems. The value obtained will be in terms of either numbers of groups per square kilometer or individuals per hectare, depending on the measurement units employed or calculated.

THE EXERCISE

The purpose of this exercise is to develop an appreciation of the difficulties of censusing primates under natural conditions and to determine an estimate of the primate population density at the field school site. Proceed as follows:

A. Determine whether the field site has established transect lines, or if it will be necessary to lay out your own lines. (Plastic flagging tape is probably the most appropriate material to use for marking transects.)

B. Design the transect layout and verify that it is feasible (i.e., it must not run over a cliff or through a lake, or over other impassible terrain). The transect length should be a minimum of one kilometer, but five kilometers, if available, is better.

C. Walk the transect and collect data as detailed in the exercise discussion. It is usually desirable to walk the transect several times and average the data. Sometimes an afternoon walk in addition to a morning walk may be possible.

D. Inspect the data and determine a value for w. This will, in most forested environments, be less than 50 meters, and often less than 20. Values above these should be viewed with suspicion and perhaps accorded the status of outliers.

E. Perform the calculations to derive an estimate of the primate populations, both groups and individuals per area measure, for each species encountered in the area.

F. Produce a final report for the exercise.

Exercise 15

A Phenological Transect Survey

What Plants Are in Which Phases of Their Cycles in the Area?

Similar in pattern to a line transect survey of populations is a phenological transect survey, though obviously the methods and objectives differ significantly. Vegetational cyclicity, the pattern of change in leaf and reproductive components, is a major force influencing many of the population dynamic and demographic factors in a primate society. Consequently, these cyclical patterns are of interest to primatologists. Vegetational cycles in the form of "stable" and "fluctuating" patterns are often labeled as "predictable" and "unpredictable" and used to evaluate habitat type (Colwell, 1974; Stearns, 1981). The importance of the plant periodicity to the ecological relationships of the primates cannot be overstated. Frankie, Baker, and Opler (1974a) make the following statement in a comparative forest study, which is perfectly applicable to primate research:

> Major patterns of leaf flushing are informative since numerous phytophagous organisms must be synchronized in their life cycles with the succulent leaf resources that become available during peak periods of leafing. . . . Conversely, major periods of leaf fall must be important to the dynamics of litter organisms that occur in both forests.
>
> Many phenological patterns observed . . . fit well with the seasonal cycle of climatic conditions. However, some patterns may be importantly influenced by biotic as well as by abiotic factors. (p. 906)

These patterns of periodicity in leaf production and leaf fall can be seen to have a straightforward relationship to the diet available to leaf-consuming species whether they are specialists on new or old leaves, but they are equally important to species that feed upon insects. The insect population will largely fall into the "numerous phytophagous organisms" category used above and their availability will be dependent upon the leaf flush. Flowering and fruiting cycles can be viewed in exactly the same fashion.

Basically, phenology then, is the study of the cyclic patterns of leafing, flowering, and fruiting. A sequence of phenophases in both leaf and reproductive components can be set up as follows, with recording codes L0 to L11 and R0 to R16, though this is a much finer set of distinctions than may be generally in use:

Leaf		**Reproductives**	
L0	dormant	R0	dormant
L1	initial bud	R1	initial bud
L2	developing bud	R2	developing bud
L3	mature bud	R3	mature bud
L4	initial leaf	R4	initial flower
L5	new leaf	R5	juvenile flower
L6	juvenile leaf	R6	mature flower
L7	full new leaf	R7	open flower
L8	mature leaf	R8	pollinated flower
L9	old leaf	R9	petals discarded
L10	degenerate leaf	R10	small immature fruit (seed set)
L11	discarded leaf	R11	juvenile fruit
		R12	full immature fruit
		R13	staged ripening of fruit—1, 2, 3, 4
		R14	fully ripe fruit
		R15	degenerate fruit
		R16	discarded fruit

Many botanical studies actually employ a limited set of phenophase categories such as: "flower buds, flowers, immature fruits, and mature fruits" (de Lampe, Bergeron, McNeil, and Leduc, 1992), but these may not be sufficiently fine grained for a primate dietary study. A similar limited set for leaf production might include only "shoots, flush or young leaves, and mature leaves" (NRC, 1981).

Methods for Phenological Study

Phenological studies are normally conducted either along a set of transect lines or through an established study plot of several hundred square meters. Generally for the purposes of a primatological study, using a transect line, selecting all trees over 4 meters in height and within 5 meters on each side of the line will produce an adequate sampling. Chapman, Wrangham, and

Chapman (1994) used such a technique in the Kibale Forest, randomly selecting 200-meter-long segments of the trail for regular phenological recording. Each of the trees within the census strip needs to be identified (usually through a number tag) and measured for diameter at breast height (DBH), height (H), and height of first branch (HFB). In some forests and for some species, it may be necessary to attempt to determine the diameter above the buttresses of trees. This can be difficult but is generally not insolvable. Once the preliminary survey data is established, the phenological data can be collected regularly during a transect walk.

On a phenological transect walk, the data recorded for each tree can be collected in several distinct fashions. It may be that the research design of the project requires only a minimal set of phenological information, and listing trees as "D" (dormant), "Sh" (shoots), "F" (flush of young leaves), "Ml" (mature leaf), "Ol" (old leaf), "DR" (dormant reproductive), "fb" (flower buds), "flo" (flowers), "If" (immature fruit), and "Mf" (mature fruit) will be sufficient. Alternatively if more detail is required, the listing and codes from the previous page may be suitable. It is also possible to make the record even more fine grained by dividing each tree into quadrants using the cardinal directions, and recording phenophases for each of the quadrants, then averaging them for the tree's final score.

One important addition to the record when accurate data on fruit production and estimates of remaining standing crop are required is a process of counting the mature fruits on the tree. Normally it would be impossible to perform a complete count, but by viewing the crown as a grid with a dozen or more sectors and by counting two or preferably three of these, it is possible to produce a reasonably close estimate of the amount of fruit present at one point in time. Sample record sheets for a basic and a fine grain phenological survey are included at the end of this exercise and are available on the accompanying CD.

Analysis

Analysis of phenological data should be done in a diachronic mode such that measures of time between first leaf, first flower, and first mature fruit for individual species can be determined. This is simple for deciduous trees that undergo some period of dormancy, but evergreen species may have to be evaluated with "leaf traps" under them and counts or weights of leaf fall evaluated on a weekly basis. For evergreen species the most obvious diachronic factor is the time between first flowers and first fruits. A diachronic approach will also yield annual cycle graphs that show the numbers of species in leaf flush, flower bud, mature fruit, and so forth (see de Lampe et al., 1992; Frankie et al., 1974a). In a short field study, such a wealth of diachronic data cannot be obtained, but sufficient data may be available to present graphic results over a three- or four-week period.

It is more likely that a synchronic study will be the norm, and cross-sectional data presentations of percentage or number of individuals in each phenological stage would be expected.

For more complex analyses where several years' worth of phenological records are available, the production of Colwell's *P, C,* and *M* indices is generally possible (Colwell, 1974; Stearns, 1981). The mathematics of calculating these indices are not simple; graduate and professional researchers are referred to the original articles for an explanation of the procedures and attendant cautions.

THE EXERCISE

The purpose of this exercise is to develop an understanding of phenology and vegetational censusing under natural conditions, and to develop a phenological record for the field school site. Proceed as follows:

A. Determine whether the field site has established transect lines or if it will be necessary to lay out your own lines. (Plastic flagging tape or paint is probably the most appropriate material to use for marking transects.)

B. Design the transect layout and verify that it is feasible (i.e., it must not run over a cliff or through a lake, or over other impassible terrain). The transect length can be variable, but it should be long enough to have a good probability of sampling most of the variant communities in the area; the rule is usually "the longer the better."

C. With the aid of a knowledgeable botanical assistant, each of the sample trees should be numbered, identified, and recorded in the site logbook. This information may already be available for the study area.

D. Walk the transect and collect data as detailed in the exercise discussion and required by your instructor. The instructor may set either a normal botanical phenology survey or a more intensive version that would be suited for a dietary study of primates. It is usually desirable to walk the transect several times during the course, in order to allow for some phenological progression to occur.

E. Inspect the data and calculate the numbers of species and individual trees in each state of the phenological cycle. If repeated surveys are done, can aspects of the cycles for different species be demonstrated?

F. Produce a final report for the exercise, taking care to complete the analyses specified by the instructor.

Basic Phenology Sampling Work Sheet

Observer:_______________________________ Date:____________________________

Site:____________________________________

Leaf Codes: D (dormant), Sh (shoots), Fl (flush of young leaves), Ml (mature leaf), Ol (old leaf)

Fruit/Flower Codes: DR (dormant reproductive), Fb (flower buds), Fl (flowers), If (immature fruit), Mf (mature fruit)

Tree	Location	Leaf Status	Fruit Status	Notes

Advanced Phenology Work Sheet

Observer:______________________________Date:____________________________

Site:___________________________________

See the following page for leaf and reproductive codes.

Tree	Location	Leaf Quadrant Code	Fruit Quadrant Code	Fruit Quadrant Counts	Notes

Code Sheet for Fine-Grained Phenology

Leaf

L0 dormant
L1 initial bud
L2 developing bud
L3 mature bud
L4 initial leaf
L5 new leaf
L6 juvenile leaf
L7 full new leaf
L8 mature leaf
L9 old leaf
L10 degenerate leaf
L11 discarded leaf

Reproductives

R0 dormant
R1 initial bud
R2 developing bud
R3 mature bud
R4 initial flower
R5 juvenile flower
R6 mature flower
R7 open flower
R8 pollinated flower
R9 petals discarded
R10 small immature fruit (seed set)
R11 juvenile fruit
R12 full immature fruit
R13 staged ripening of fruit—1, 2, 3, 4
R14 fully ripe fruit
R15 degenerate fruit
R16 discarded fruit

Exercise 16

A Home Range Survey
What Is the Group's Home Range? Quadrats versus Least Polygon

Determination of the area that a primate individual or group ranges over is a basic data-collection exercise that can become extremely complex in analysis. To begin with, there are three related terms commonly employed in primatological study: "annual home range"; "territory"; and "core area." It is worth emphasizing that two of these are observational artifacts, and only one of them is present in the mind of the primate involved. This latter feature is territory—most commonly defined as an area reserved to the exclusive use of an individual or group and usually defended against conspecifics. Thus, the territory may be perceived by the primate or primates as a possession to be defended. The other two concepts are arbitrary constructions of the observer. They are bound by a fixed time span in the first, and by an arbitrary "majority (or percentage) time spent in area" statistic in the second. Both are human creations and might not be recognized by the primates. It is also worth noting that home ranges are almost always larger than territories, which, in turn, are larger than core areas.

This exercise centers around the techniques used to determine a home range for a subject species, though it will not be an annual range. Home ranges can be calculated for lesser periods, and ten-day ranges were frequently used in early studies, as have been monthly, quarterly, and seasonal range collections. Lehner (1996) points out that ranges can be obtained through two methods: (1) direct observation, and (2) indirect methods, which include: tracks, capture-recapture, radioactive tagging, baits with ingestion dyes to color urine or feces, surveillance cameras, and radio-telemetry. We will deal with only the direct observation process, as the indirect methods are not normally available or useful to primatologists.

The simplest methodology involves recording the locations of observed individuals or groups on a map plot of the study locality. This is easiest in an area that has been gridded with transect trails. Often these lines will be placed at 100-meter intervals, providing a set of one-hectare-sized blocks and making calculation of the range straightforward. If continuous observations are taken, and complete day ranges obtained, a simple process of plotting the set of day ranges on the map grid will enable the researcher to draw a line around the area delineating the "home" range. This can be done in two ways: (1) through counting the number of hectare blocks in which the subjects are seen, or (2) by drawing a "least polygon" around the areas used, and counting quarter- or half-hectare units where the polygon line crosses a block.

Figure 16.1 shows the difference between these two processes for a small grid.

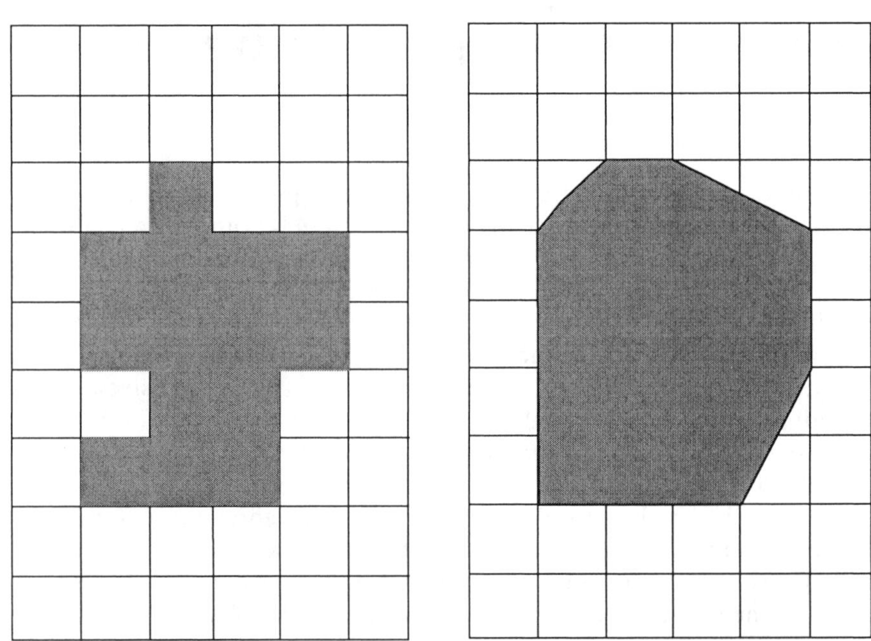

Figure 16.1 Two methods of delineating home range. A presents a block count of 14 hectares, while B, the least polygon, actually encloses a larger area of 17.5 hectares.

While this may seem to be a simple proposition in either case, and it initially appears that the polygon includes an error that enlarges the area, these are not critical features, nor do they merely represent minimum and maximum estimates of the true home range. Varying the number of perimeter points can dramatically change the size of the area enclosed in a polygon, but it also increases the difficulty in constructing it.

A relevant technique developed by Odum and Kuenzler (1955) for ornithology is based on the principle of diminishing returns. They assume that as the numbers of observations increase, the rate of increase in area will decrease with each increment. This means that the measured home range area reaches toward an asymptotic approach to the "real" home range area, and consequently and arbitrarily they applied a 1 percent rule. This is the point on the curve at which each added observation produces an increment of less than 1 percent of the area calculated as the final size. The purpose of this technique is not to estimate range size but to estimate the number of observations necessary to validly calculate the range area. The technique is much easier to apply to birds than mammals, where it has been estimated that radio telemetry would be the appropriate observational method. A discussion of the Odum and Kuenzler technique as well as the indirect methods can be found in Lehner (1996:510–521).

Computer programs that go well beyond the limitations of this basic process are available for the Macintosh. Antelope 1.3 and Home Range 2.2 can perform the basic calculations for this exercise and go on to provide an impressive array of additional and very sophisticated analyses. Investigation of these capabilities is encouraged. As of March 2000, these programs were available from the Web sites of the authors: http://www-biology.ucsd.edu/research/vehrenbury/programs.html will provide Antelope 1.3 and The Kernel 1.0 (a simple program that uses the kernel method to estimate home ranges). Home Range 2.2 can be found at http://caspar.bgsu.edu/~software/Analysis.html. All are interesting and useful adjuncts to any analysis of home range and territory data.

The Exercise

The purpose of this exercise is to calculate short-duration home range areas during the performance of a field school practical study. The main operative assumption is that the exercise would not be conducted as a single project, but would be coincident with other projects.

A. It is assumed that mapping and gridding of the study area are available or that transect lines have been marked in some fashion (see exercises 14 and 15).

B. With a topographical map of the study area and a clear plastic transparency sheet, draw out the transect lines and create a 100 meter by 100 meter grid overlay. (This assumes that the site employs a 100 x 100 grid; it may be 50 x 50 or 200 x 200 meters. Adjust as necessary to complete the exercise.)

C. From the observational data mark the locations at which the study subjects have been observed. It may be appropriate to distinguish between levels of use for each hectare by placing a count of encounters, or of hours observed, in the block.

D. Employing the two techniques illustrated above—a simple count of hectares, and a least polygon—estimate the home range area as you have observed it. An important query may be answered from your data: Will an increase in the number of points used in the construction of a polygon improve the estimate of range size? Perhaps you should calculate the area with ten, twenty, and thirty perimeter points.

E. Prepare a report in the format and with any further analyses as specified by the instructor.

Exercise 17

Vegetation Sample Plots and Indices

Any study of ecological relationships in primatology will eventually require the development of a vegetational plot or a comparative set of plots. There are a number of indices of ecological complexity, usually in the form of an index or coefficient of dispersion, that can be applied to vegetation plots and yield useful information applicable to problems in primate behavior and socioecology. These indices of dispersion can also be used to measure the spread of animals throughout the environment, and at a fine-grained analysis, they can assist in studies of the temporal patterns of habitat usage.

The most basic and useful calculations for the distribution of animal and plant species within the community are Morisita's index of dispersion and the standardized Morisita index. For comparing the species composition of sample plots three calculations provide measures of similarity. The first is Morisita's index of similarity, the second is Morisita's simplified index of similarity, and the third is percentage similarity (also known as the Renkonen index). This exercise will focus on the three calculations of index of dispersion (I_d), index of similarity (C_λ), and percentage similarity (P).

Preliminaries

All of these indices presume the existence of a set of quadrats or sample plots for which there is some form of frequency data. The most common type of data is a count of number of individuals of a group (for an estimate of group dispersion) or of species present (for estimates of the dispersion pattern or vegetation diversity). Thus, two models of data collection can be proposed. The first, suited to a study of group dispersion, involves counting the number of individuals present in quadrats of a grid, preferably for a particular point in time. This might be performed by a team walking a set of grid lines and recording all individuals of a species seen, keeping track of the midline of the quadrat so as not to double count subjects. Once the data is assembled, a representative grid with frequency counts is the basis for the calculation of I_d. The second model is appropriate to vegetative sampling and is often referred to as a sample plot. Sample plots can vary greatly depending upon the particular investigation, but generally a square or rectangular area is marked off and data collected about all of the individual plants within the quadrat. As with many other forms of sampling, these plots should form a representative sample of the whole area. While it is typical for individuals in such plots to be measured for DBH, height, crown volume, height of first branch, and so forth, the data required for the indices described here are only counts of identified species.

Morisita's Index of Dispersion

Morisita, in 1962, presented an index of dispersion that can be applied to individual species (both plant and animal) and provides an index that is independent of population density, though it is affected by sample size. It does have a known sampling distribution, and hence, a version of the chi-square test can be employed to validate the results (see Krebs, 1999:216). The formula for I_d is:

$$I_d = n\left(\frac{\Sigma x^2 - \Sigma x}{(\Sigma x)^2 - \Sigma x}\right)$$

nd where n is sample size, the number of quadrats measured
Σ x is the sum of the counts in each quadrat
Σ x^2 is the sum of quadrat counts squared.

These values can be most easily calculated and organized in the same fashion as for a standard deviation (illustrated on p. 53).

To operationalize the calculation, one must have laid out a set of quadrats and counted members of individual species present in each quadrat. For the practical purposes of an exercise this might be constrained to 10-, 20- or 25-meter square quadrats, and include only the main canopy trees, or those over a particular DBH. In primatological studies, two criteria are common: use of a DBH greater than 4 centimeters or use of all trees taller than 4 meters. Then it is only necessary to make a two-column table, one for the x values and one for the square of x, adding the columns then produces the summation values to insert into the formula. Alternatively, an Excel worksheet can be found on the CD; this in conjunction with the Excel program will do the calculation.

MORISITA'S INDEX OF SIMILARITY

Morisita in 1959 proposed this index to measure similarity between two communities or sample sets. Students are advised *not* to confuse this with his index of dispersion. They are very distinct items, and evaluate different data. The index is traditionally represented as C_λ but has occasionally appeared in the literature as I_s. In either case, it can be interpreted as a probability statement. The probability that an individual organism drawn from one sample, and one drawn from a second sample will be of the same species, divided by the probability that two individuals drawn from either sample one or two will be of the same species, is the index. The formula is moderately complicated:

$$C_\lambda = \frac{2 \cdot \Sigma X_{i1} X_{i2}}{(\lambda_1 + \lambda_2) N_1 N_2}$$

and where X_{i1} = Number of individuals of species i in sample one.
X_{i2} = Number of individuals of species i in sample two.
N_1 = Total number of individuals (all species) in sample one.
N_2 = Total number of individuals (all species) in sample two.

$$\lambda_1 = \frac{\Sigma(X_{i1} \cdot (X_{i1} - 1))}{N_1(N_1 - 1)}$$

and

$$\lambda_2 = \frac{\Sigma(X_{i2} \cdot (X_{i2} - 1))}{N_2(N_2 - 1)}$$

The implementation is like that for the index of dispersion, simply a matter of counting individuals of identified species in two sample quadrats, and placing the counts into the formula. Note that this is a calculation for only one species and would be repeated for each additional species. An overall estimate or index of similarity between two sample plots is available as an Excel worksheet on the CD. The data needed is a listing of species and counts of individuals for two sample plots, and execution is a matter of typing in the appropriate numbers—note that the same line should be used for a species in each sample—and the calculation is automatic.

Morisita's index of similarity is nearly independent of sample size, is exceedingly robust, and has been recommended as the best overall measure of similarity for ecological use.

PERCENTAGE SIMILARITY (RENKONEN INDEX)

The Renkonen index (Renkonen, 1938) requires the standardization of each community sample as percentages for each species found. These can be checked easily as they should sum to 100 percent. The formula requires the calculation of a sum of the minimum values and requires an illustration of the procedure to clearly understand the equation. The equation is:

$$P = \Sigma \text{ minimum}(p_{1i}, p_{2i})$$

where P = Percentage similarity between sample 1 and sample 2.
p_{1i} = Percentage of species i in sample 1.
p_{2i} = Percentage of species i in sample 2.

The calculation is a simple one, but it is important to recognize that what is required is the "choice" of the minimum value for a species in either sample one or sample two. Thus if for three species we find the following percentages in two sample plots:

Species	Plot 1	Plot 2
A	36.5	1.7
B	11.0	47.2
C	2.9	3.1

then in our summation process we select the minimum value of the two for each species and get 1.7 for *A*, 11.0 for *B*, and 2.9 for *C*. The resultant sum is a percentage value for the similarity between the two plots. The calculation is very simple and an excellent quantitative measure. Note that this is only a partial calculation since the percentage total for each plot must add to 100%. No worksheet is provided for this calculation.

THE EXERCISE

This exercise is to provide practice in the use of ecological indices of similarity and evaluate the dispersion of a species. Proceed as follows:

A. Determine whether the field site has established sample plots or if it will necessitate laying out your own.

B. Design the sample plot layout and verify that it can be examined in the time available. As suggested, 10-, 20-, or 25-meter plots and a minimum DBH or height measure may be imposed by the instructor.

C. With the aid of a knowledgeable botanical assistant, each of the sample trees should be numbered, identified, and recorded in the site logbook. This may already be available for the sample plot.

D. Inspect the data and calculate the index of dispersion for the most common species and for the least common. Calculate the index of similarity and the percentage similarity for the sample plots.

E. Produce a final report for the exercise, taking care to complete the analyses specified by the instructor.

Additional Indices

There are a large number of other ecological data indices in the literature, and it would be very worthwhile to consult Krebs's *Ecological Methodology* (1999) or Pileou's *Ecological Diversity* (1975) for coverage of them.

However, the CD also provides worksheets for three indices often used in primate studies:

Rasmussen's **RU**, a two-sample test of diversity and distance between quadrats;

Stander's **index of diet diversity**, and

Stander's **correlational index**.

The worksheets are largely self-informative and versions for Excel 3.0 and for Excel 5.0 or later are present. The data required for Rasmussen's RU are the number of individuals of each species in each quadrat, and the distance between the centers of each pair of quadrats. The Stander calculations require a species list and the relative proportion of that species in the diet of the sample population. Stander's correlational index requires this data for both populations. The worksheets on the CD may have more indices available; please consult the worksheet Read Me file.

Exercise 18

Vegetative Sampling and Voucher Specimens

One of the normative activities for any field study is, or at least should be, the collection of plant samples for the purposes of identification and, secondarily, for the maintenance of a herbarium collection. This exercise is not intended to fulfill all of these objectives but merely to provide experiential learning in the processes and practices of collecting adequate samples. By an "adequate" sample, it is implied that the amount of material and the extent of the structures sampled are sufficient to enable a herbarium technician to identify it to the species level. Should a researcher collect a plant new to science, sufficient material is needed to provide the requisite "voucher specimens" that will serve as the future "type" specimens. These two steps are normally sequential with the second triggered by a response of "unidentified/new plant" from the herbarium. Lest you think this unlikely, I should mention that in 1996 the Budongo Forest Project forwarded two new species to the botanical literature from a thoroughly explored forest that has been intensively studied and under managed logging since the 1920s (V. Reynolds, personal communication, October 1996).

Plant Specimen Sampling

You have just seen a primate consume the leaves of a plant, and your research assistant cannot identify the species—what do you do now? Obviously, you want to collect a specimen from the plant and have it identified at a herbarium. How much material do you need to collect? One botanist of my acquaintance said, perhaps facetiously, ". . . the whole thing!" Generally speaking, the person conducting the identification would like to have too much rather than too little material. It is usually appropriate to attempt to collect not a single leaf, but a whole leaf cluster; not a single flower, but a whole flowering mass, and so forth. For small plants, an excellent sample would include the leaves, flowers, and fruit (in season), and for some varieties, the root, rhizome, or tuber. A leaf cluster may be the minimum acceptable sample. Once the sample is collected, it is likely that it cannot be immediately presented to the botanical technician but must be preserved in a manner that retains as much information as possible. The normal process utilized is drying in a plant press.

A plant press consists of four components: two outside plates, ventilators, dryers (blotting paper or felt), and a mechanism to apply pressure to the stack. Commercially manufactured presses are available at modest cost, but a perfectly adequate one can be constructed of two pieces of plywood (roughly 30 by 45 cm or 12 by 18 inches), two webbing straps (with buckles), and a pile of corrugated cardboard in the same size as the outer boards, blotters, and newspaper. The standard pressing process involves placement of the sample carefully between sheets of newspaper, a blotter and cardboard on each side of the specimen. A number of specimen sets are stacked between the pressboards. The web straps are then used to apply a strong compression to the stack. The loaded press is then placed on edge, with the cardboard corrugations aligned vertically, in a location that provides a flow of warm dry air. In the field this may be in the rafters of the field hut, on a rack beside the cooking fire, or suspended at a safe distance above a gas lamp. It is advisable to replace the dryers and ventilators at least once per day in order to avoid the development of mold on the specimens, as well as to speed up the process. Some plant materials will adhere strongly to paper, and others, particularly large succulent or fleshy plant materials, cannot be dried satisfactorily without cutting the specimen into appropriate forms. In the former case, directly drying the specimen on a mounting sheet with a layer of wax paper may suffice; in the latter, cutting the specimen so as to create an outside layer that can be spread out flat is the common procedure. Alternatively, succulents and fleshy plant materials can be submerged in sample bottles under formal-acetic-alcohol (FAA), which will also preserve the original colors of the specimen.

FAA (formal-acetic-alcohol)

Combine equal parts water and 95 percent ethanol (may or may not be "denatured"). Add formaldehyde (40 percent) and acetic acid (glacial—35 percent) each in the amount of 2 percent of the volume of the water-alcohol mixture. Preferably store in dark bottles.

Field Notes

Whenever specimens are collected, include accurate information as to the locale, longitude and latitude, the surroundings of the specimen—in particular, the plant community type, associated plants, altitude, slope of the ground, nature of the soil, its moisture content, and other pertinent information.

For the specimen itself, the observer should record the average size of specimens, the color and pattern of the flower, the nature of the bark (for trees), and color information in case the specimen fades during drying. (A standard color card or a photograph with such a card next to the specimen is often useful and appropriate.)

Mounting—Voucher Specimens

Standard mounting paper for herbarium use is a sulfur-free heavy paper of 30 by 40 cm (11.5 by 16.5 inch) size. The specimen may be folded to fit onto the sheet and can be attached with special herbarium tape, a diluted white-glue mixture, or one of the newer herbarium fixative sprays. In any case, the dried plant is affixed to the paper, the field note data is transferred directly to the sheet, and, if necessary, an envelope for loose components is attached. This entire sheet constitutes the voucher specimen for the plant.

The Exercise

The purpose of this exercise is to provide practice in the collection, preparation, and mounting of plant specimens relevant to a behavioral ecology study of a primate species.

A. Prepare a listing of specimens to collect (it might be most appropriate to sample the species that are fed upon by the local primates).

B. Proceed to collect samples of the species on your list. Make certain the sample is large enough to provide a complete leaf cluster; add fruit and/or flowers if available; and prepare field notes.

C. With the aid of a knowledgeable botanical assistant, each of the samples should be numbered, identified, and recorded in field notes and logbooks.

D. Place specimens into the plant press and dry them for several days. Using appropriate mounting media, affix the specimens to herbarium pages and transfer the field note data.

E. Produce a final report for the exercise as specified by the instructor, and submit it with the herbarium samples.

Exercise 19

Fecal Analysis
Washing and Examination for Diet

"Oh Yuck!"—the common student response to this subject, even at the graduate level.

Unfortunately, the analysis of fecal material is a mainstay of primate behavioral ecology and an important mechanism to assess the dietary pattern of primates. It is also important for understanding the relationships of primates to their food species, especially when questions of seed dispersal and seed predation are crucial to the research design.

One tenet of behavioral ecology studies is that primates tend to be "wasteful" or "highly selective" feeders, depending upon the source of the statement. It is well recognized that primates may not ingest a significant portion of what they remove from plants, and patterns of discard are important data in all dietary studies. Primates often do not ingest all of what they place in their mouths and chew. Hence, study of the practice of "seed spitting" has recently been in vogue.

As in any observational work, data collection routines that focus upon selection, discard, seed spitting, and so forth, are susceptible to observer bias and error. A half century ago, the solution to this was direct examination of stomach contents of shot specimens. At present, this practice is considered to be unethical and is usually illegal in most countries. Other techniques must be used. The simplest and most direct is examination of the fecal output. While this has limitations in that some of the material ingested will not leave remnants that are identifiable even under microscopic examination, feces may contain a variety of intestinal parasites, their eggs, egg cases, larvae, or cysts that are of interest to the primatologist since they provide evidence of additional ecological cycles and relationships.

Seed dispersal involves the concept of a symbiotic feedback loop between plant and animal. The plant provides the primate with some form of food material, and the primate, in its daily activities, transports and deposits seeds away from the parent plant. Establishing a proof or verification of this relationship is not easy. A long-standing experimental procedure used in support of seed dispersion studies involves extracting the seeds from a stool and planting them in germination trays, or placing them between sheets of wet material. One then calculates the proportion of seeds that actually germinate.

Seed predation, in comparison, is understudied. Seed predation models suggest that some portion, or all, of the seeds ingested by a primate may be crushed, broken, or chemically digested during their passage through the gut. As a consequence, the seeds are destroyed and the plant does not receive any benefit from the primate.

The dispersion and destruction of seeds is not a simple either-or situation; for many plants and primates, some intermediate condition exists. The primate may be acting as both seed disperser and seed predator. The use of fecal analysis in studies of seed dispersion and predation may be the most direct method of examining the relationship between the two organisms.

The technique pioneered by Overdorff and Strait (1996) can be employed to clarify the role of seed consumption for a species, especially when combined with observations of feeding behavior (Paterson, 1997). The procedure involves washing and drying of a fecal sample, followed by examination of the residue at low power under a microscope. The examination is to sort those seeds that are undamaged from those that have had the seed casing scratched, broken, or digested. Once the numbers of intact versus damaged seeds have been determined, a proportionate index of seed predation for each species of plant can be calculated. Such an investigation cannot provide a positive statement about seed dispersal, but it certainly can yield information about the seed predatory actions of a primate. After sorting, undamaged seeds can then be carried into a germination experiment; preliminary sorting provides a more valid test of germination potential than the use of unsorted seed.

An extremely useful resource for this type of research is a seed catalog for the field site, and I urge that such a collection be part of the base data collected for each and every study locale.

The Exercise

This exercise is intended to provide practice in the technique and procedures of fecal sample analysis. The experimental equipment required are: small plastic sample bags; a set of standard screens, or at least one screen of 80 or 100 mesh (if these are not available, a suitable substitute is a fine mesh stainless steel tea strainer—carefully marked as to its use!); a source of clean water—allowance of approximately 1.5 liters for each sample is appropriate;

surgical/general purpose latex or rubber gloves; a sorting board—easily made by marking a grid onto a flat board or plastic slab; and a low-power dissecting or stereo microscope. A 10 or 20 power hand lens may be substituted if a microscope is not available.

A. During the course of observational studies take careful note of the location of any falling feces. Between samples, carefully move to find the fecal material. A labeled sample bag can be turned partially inside out and the sample collected without contamination, and sealed.

B. Wearing gloves, conduct the washing of the sample in a screen of 80 or 100 mesh. Most of the soluble mass will be washed away, and a residue of plant fiber, seeds, wood, bark, rock, insect exoskeleton, and so forth, will remain.

C. Turn out the washed material onto a sheet of plain paper (or preferably filter or blotting paper) and set out to dry. At this time it may be possible to find larvae and tapeworm segments or egg cases in the sample. These should be removed immediately and stored in a vial of alcohol or FAA (see exercise 18 for the formula).

D. After one or two days of drying (longer if the material is to be carried home for final analysis), the sorting board, microscope (or hand lens), and, if available, the seed catalog are used to sort and identify the damaged and undamaged seeds.

E. Calculate the relative proportion of damaged seeds for each species of food, and produce a report to the specifications of the course instructor.

Applied Study Projects

Branching Out into Research Design

The following four exercises are intended as projects for the development of research designs and/or as small research projects in themselves. Each presents a research problem and makes some suggestions as to how it may be organized, but the development of an acceptable design and protocols is the exercise.

Exercise 20

A Spatial Location Study

(The content of this exercise was contributed by Dr. Sue Taylor Parker, Anthropology, Sonoma State University, as a component in a larger full-term observational project, and it has been somewhat modified from the original form.)

One relevant question for any primate study is: "Do the individual animals move evenly through their environments, or do they have preferred usage sites and pathways?" Intuitively, since we recognize such patterns in our own lifestyles, we would posit that they do. However, as scientists we would like to be able to show support for this conclusion, as well as investigate the patterns that may be employed by individuals, by different sex and age groups, and so forth. A large number of possibilities are open to development once the basic concept of locational studies has been selected.

This exercise is straightforward and can be carried out with primate populations in zoological enclosures. The first operation after selection of the study group is to "map" the cage/enclosure areas. The map may be a simple gridding of the enclosure floor into quadrants (just the four quarters) or six, eight, nine, twelve, or sixteen quadrats. In more complex enclosures, particularly those that offer substantial use of the third dimension, a set of cubic measures would be more appropriate. It is unlikely that the facility will be amenable to marking these divisions directly on the enclosure, so the most appropriate technique is to create a map that uses clearly observable features to establish boundaries between quadrats. Observational techniques such as focal animal sampling could be used, but the interval process of a group-oriented scan sampling can be used to collect large amounts of data in a relatively short period.

Once the protocol is established, the mapping done, and the data-recording technique selected, some consideration of the analytical issues should be included, *before* the start of the collecting phase. The data collected from this study, at least in the design pattern suggested here, is *not* amenable to simple statistical testing.

The most appropriate analysis where there are data represented as frequency counts in a large number of cells, whether they are arranged as a linear or a row and column table, is the G-test. The G-test or G-statistic is perhaps more typically found in social science statistics texts as the χ^2 (chi-square) table test. An important consideration for employment of this statistic is the interpretation of the probability value derived. If the value of G reaches the critical level (still the < 0.05 criterium), this signifies rejection of the *null* hypothesis, which always is: *that there is no effect (of location) in the data*. An appropriate interpretation is that the alternative hypothesis (that there *is* an effect of location) may be accepted. This is one of the circumstances in which the test process is interpreted in a different fashion to the common chi-square test. It is possible to test individual cells against each other using the regular chi-square, but with a large number of cells the number of tests to be run increases rapidly.

With these cautions in mind, the student is invited to construct a research design for a spatial locational study. The three critical components are:

1. Construction of an appropriate cellular division of the study environment
2. Selection of an appropriate observational protocol and
3. A statement of the analytical processes to be employed

Once the research proposal is approved by your instructor, you might be directed to seek approval of the project from the appropriate ACC and to carry out the study, producing a standard scientific report.

Exercise 21

Exploring Postural Congruence

Have you ever noticed individuals in reasonable proximity to or interacting with each other, employing the same postural positions? This phenomenon is commonly present in herds, flocks, and schools of animals and is most often seen as coordinated movement. It has, however, been ignored as a behavior pattern in individualistic species such as primates. Sharing of postures is such a common human practice that thirty-five years ago A. E. Scheflen (1964, 1965, 1972) proposed that such interactions should be called "postural congruence" and be treated as a communicative adjunct to speech. Scheflen went on to suggest that the pattern of postural sequencing was linked to the particular language used by each individual. He was suggesting that the patterns were culturally conditioned. Until recently this study of nonverbal signaling has been employed only in psychiatry and psychology. It was assumed by investigators that nonhuman primates would not display these types of signals since they could not speak. Recent studies by Boyd (1998) and by Jazrawi (2000) have confirmed that postural congruence functions as part of primate communication.

Postural congruence, abbreviated here to "PC," has also been called "postural echo" (Beattie and Beattie, 1981). PC can be grouped under the more general pattern of "model and mimic" in that one individual adopts a posture (becomes the model), and a second individual subsequently moves into the same posture (is the mimic). PC can be normal or parallel with both individuals orientated in the same direction, or in a mirror-image form that most commonly occurs with the two subjects facing each other.

The psychiatry-psychology literature argues that the use of PC is related either to an attempt at personal gain by an individual or to signals that indicate agreement or solidarity between individuals. PC might be used by high-ranking members of a hierarchy to indicate unity in the face of challenges, or by individuals intent upon social manipulation to attempt to rise in rank, or to obtain access to some resource controlled by another. In other words, interpretation of the behavior is very context specific. Both baboons and chimpanzees waiting to secure a morsel of meat from a prey carcass adopt the posture of the "possessor" of the carcass. Females wanting to touch or play with a newborn adopt the same posture as the mother in order to get closer. There are innumerable instances of recorded postural congruence, but it has only recently become a subject of study.

PC may be divided into "complete" or whole body congruence, and "partial" or half-body congruence where only the upper *or* lower body and limbs are in the same postures. PC may also occur in differing patterns associated with age and sex categories, and with rank in a hierarchy. There is also some suggestion that PC differs in pattern of use between female-bonded and non–female-bonded social systems.

The exercise is to research the literature and construct a research design to study the role or effect of PC in a social group. The design, as always, requires:

- Clear and careful defining of terms
- Construction of a useful ethogram and a "posturogram"
- Detailing of the variables to be recorded, and
- The manner in which analysis will be conducted

Care needs to be taken in setting the observational "boundary conditions," such as: What is the maximum distance at which two individuals can be considered as still "interacting" and not just being in congruence by chance? What is the chance probability that two individuals are in postural congruence because they are performing the same behavior? These are just two of the possible confounding situations in this type of study.

Your instructor may provide a more detailed set of requirements for the completion of the proposal document. Conduct of the study after approval by an ACC may also be the major focus of a senior or graduate exercise.

Exercise 22

A Nearest-Neighbor Study

Do social primates form regular or consistent subgroupings within a group? Do these subgroupings, assuming they exist, change during the course of a daily cycle or in relation to differing activities? Are subgroups determined by rank, or are they a consequence of rank seeking? There are enormous numbers of questions in this vein that relate to the patterns of interaction within primate societies, and many possible ways to study them. One method that is appropriate and straightforward is a nearest-neighbor study. The basic premise for studies of subgroup structure is that individuals can be found in associations because they prefer to be with some group members and avoid other individuals.

The patterns of subgrouping can be statistically analyzed only when data about identified individuals and their distances from each other are available. Nearest-neighbor data can also be used to assess the "sociability" of individual animals. If they are constantly in a cluster and in close contact, their rating would be higher on the scale than if they were rarely in any association or kept other individuals at greater distances. There are numerous research designs that can be constructed to employ nearest-neighbor data and successfully yield information about the social structure of a primate group.

Basic data collected in such a study might be the distance from a focal individual to others and the duration of time that neighbors are in different distance categories. An error frequently seen in these studies involves the assumption that the observer can judge precisely distances between individual subjects at all distances. It is inappropriate to use a scale that employs one- or half-meter intervals out to 10 or more meters between subjects. The precision is unjustified and, for all but a phenomenal few, unachievable. It is far more useful and practical to use a scale consisting of "in contact, within arm's reach, within three meters, beyond three meters." It is seldom necessary to employ a finer grain, and it is difficult to verify the observer's accuracy at distance judgment.

The exercise is to devise a research question that will be answerable through the data collected in a nearest-neighbor study, to set up the appropriate variables, measurements, and analytical mechanisms that would be used in the conduct of such an experiment. Your instructor may set additional parameters and require that an experiment be conducted in order to verify the research design.

Exercise 23

Sequential Behavior Studies

One of the least-studied aspects of primate behavior is the sequencing of behavioral units. Many of the exercises in this text have emphasized continuous recording of data, appropriate coding systems, and analysis of the rates of occurrence. But data that have been recorded under these operational parameters are also suited to sequential analysis. One of the advantages of the focal animal sampling method emphasized by Altmann (1974) is the utility of the data for the study of sequences. However, the study of behavior sequences and the analysis of sequential behavior has not been common in primatology in spite of its acknowledged value. In large part this is because the analytic methods and, particularly, the statistics employed are unfamiliar to most primatologists whether they originate from the zoology, anthropology, or psychology disciplines.

A part of the reluctance to use sequential analysis derives from the lack of readily available computer programs with which to analyze sequential data. Most primatologists and student researchers will collect data that can be examined sequentially, and it is really an unfortunate oversight that the analytic programs, now available, are not used. Bakeman and Gottman (1997) and Bakeman and Quera (1995) are the core materials relevant to sequential analysis. The latter volume also includes a diskette containing the Sequential Data Interchange Standard, Generalized Sequential Querier (SIDS-GSEQ) programs for MS-DOS, to facilitate the implementation of analysis. Sequential analysis and the programs are not detailed here, as students will find reading Bakeman and Gottman and Bakeman and Quera much more informative.

The exercise is to develop hypotheses about behavioral sequences and to establish a suite of variables and measurements that will provide data for use in the SIDS-GSEQ programs. The formation of sequential hypotheses is not a difficult proposition, but it does require a slightly different way of thinking about the data. If large datasets of primatological data are available, these hypotheses may be applied to them and secondary analyses undertaken. Your instructor, as always, has the option of requiring more or less and setting limits and parameters for the conduct of this exercise.

Appendix

Replicating Previous Findings

The following is an example of a student-written scientific report for an earlier (1980s) version of the focal animal sampling exercise, (complete with Canadian spellings and grammatical errors).

Introduction

The importance of studying captive primates has been noted by previous researchers (Hall & Goswell, 1964; Hall & Mayer, 1967). Since naturalistic (field) studies are necessarily constrained by practical problems (e.g. imposing controls in natural conditions), a more complete picture may be obtained by the complimentary study of captive primates. Hall and Mayer (1967) indicate that results from captive animal studies closely correspond to naturalistic study results, therefore the merits of captive animal study are not obscured by the unnatural conditions of captive life.

The current study shall attempt to replicate some of the major findings of Hall and Mayer (1967). (Replication is an important aspect of the scientific method allowing one to test the reliability of previous research conclusions.) Hall and Mayer found: the adult male of the troop interacted with others very little, instead remaining aloof; one female acted as a leader over the others; a pubescent male received attacks of an unprecedented nature from the adult male, and "aunt" behaviour occurred towards a newborn (by other females). These results were from a study of a two-year duration on *Erythrocebus patas*. Due to the limited scope of the current study (only seven patas monkeys and 50 minutes observation time for each subject) only two of these findings shall be investigated.

It is specifically hypothesized: that the adult male shall engage in less grooming and aggression than any other member of the troop and that a single female shall emerge as a leader (engaging in more aggression over the other females and grooming the adult male more than the others) amongst the females.

METHOD AND MATERIALS

A group of seven *Erythrocebus patas* monkeys were observed in an indoor enclosure made up of two cells of equal proportion joined by a small opening. Dimensions were 50' x 12' x 8'. An outdoor enclosure was 50' in length and depth and 20' in height. Each area had rock platforms and a small pond. The troop consisted of a single adult male, four adult females and two infants. (One adult female was pregnant and the two smaller females were lactating.)

Each of the subjects was observed by use of focal animal sampling technique. This means that over the course of five (10 minute duration) samples (for each animal) the subject was the focus of intense observation and all behaviours were recorded. Both events and states were recorded. A timepiece, pen and paper were the only materials needed. Exhaustive and mutually exclusive behaviour categories were developed based upon Hall and Mayer (1967).

Behaviour categories and definitions are:

Glare — To stare intently at another with eyes wide open and mouth slightly open.

Threat-face — Eyebrows raised, eyes wide open, mouth open and head stretched towards opponent.

Attacking run — To move quickly towards another with the head stretched out and body close to ground.

Stalk — To walk slowly towards another with stiff legs causing the other to keep constantly on the move.

Chase — High speed pursuit of another.

Lunge-grab — Aggressor lashes out at another, striking or grabbing its fur.

Fight — To grapple face to face biting each other.

Withdraw — To move away from another's glare or threat behaviours.

Grooming — To pick at, scratch or remove debris from fur of self or others.

Resting — Behaviours in which no locomotion occurs regardless of posture.

Approach — To move towards another animal in a non-threatening manner (i.e. not an attack, stalk, chase, lunge, or fight).

Out of sight (OS)	Primates not visible.
Gazing	Looking around at others or spectators.
Other	This category includes behaviours not essential to the purpose of this study. For instance; elimination, playing, locomotion such as pacing, drinking, etc.
Feeding	To take in water, or solid foods (into mouth) chew, and swallow.

Observation records were scattered throughout the day evenly for each subject. Dates of observation were June 9, 1986, 9:00 a.m.–1:00 p.m. June 13, 1986, 2:00 p.m.–6:00 p.m. Total time of observation for all animals was 350 minutes (5.8 hours).

Results

Results are summarized in Tables 1 and 2 for event and state behaviours respectively. Frequencies and durations were converted to a percent score by dividing each raw score by the total (frequency or duration) for all behaviours and multiplying by 100. Results for the group as a whole indicate low frequencies of aggressive behaviours (glare, threat-face, attacking run, stalking, chasing, lunging, fighting), withdrawal, approach and gazing. (Low defined as less than 10%). High frequency behaviours were feeding, grooming, resting, and other (greater than 10%).

Individual results indicate that the adult male engaged in grooming less than the adult females but at about the same level as the juveniles. Most of his glares, threat-face, and lunges were at spectators. The most frequent behaviours were gazing, and feeding. Resting, gazing and feeding were also of a longer duration than other members of the troop with the result that the adult male interacted less with the others.

Results for the female adult members of the troop indicate most aggressive behaviours occurred between two females (subjects 3 and 4). The remaining two females withdrew and approached less than other members of the troop. Females gazed and fed at about the same levels (less than the adult male but more than the juveniles). Adult female #4 groomed the least frequently but had the longest duration of grooming (the adult male was second in grooming duration).

Aggressive behaviours were rarely seen amongst the juveniles, but when a chase occurred, a female intercepted between the juveniles. Most of the juvenile behaviours were in the "other" category and actually were play behaviours. (Errors may exist in the data since it is often difficult to determine if play is aggressive or not.)

Most grooming by subject #3 was to the adult male, and she was within a closer proximity to him more than the other females.

Table 1 Event Data (Frequency and frequency percents)

Behaviour	Total	%	AM	%	AF	%	SAF	%	PAF	%	SAF	%	Juv	%	Juv	%
Glare	12	1.8	6	6.5	-	-	2	2.1	4	3.6	-	-	-	-	-	-
Threat-face	11	1.6	5	5.4	-	-	-	-	6	5.5	-	-	-	-	-	-
Attack, run	2	.3	-	-	-	-	1	1.0	1	.9	-	-	-	-	-	-
Stalk	4	.6	-	-	-	-	-	-	4	3.6	-	-	-	-	-	-
Chase	15	2.3	2	2.1	-	-	3	3.1	6	5.5	-	-	4	4.3	-	-
Lunge	5	.7	3	3.2	-	-	-	-	2	1.8	-	-	-	-	-	-
Fight	0	-	-	-	-	-	-	-	-	-	-	-	-	-	-	-
Withdraw	38	5.7	2	2.1	5	4.8	2	2.1	-	-	4	4.7	10	10.8	15	17
Grooming	84	12.5	5	5.4	17	16.3	17	17.5	15	13.6	19	22.1	5	5.4	6	6.8
Resting	17	17.4	5	5.4	19	18.3	23	23.7	30	27.3	21	24.4	6	6.5	13	14.3
Approach	34	5.1	1	1.0	4	3.8	7	7.2	-	-	5	5.8	8	8.6	9	10.3
OS	7	1.0	-	-	1	.9	-	-	-	-	-	-	2	2.1	2	2.2
Gazing	49	7.3	22	23.9	8	7.7	3	3.1	6	5.5	4	4.7	4	4.3	2	2.3
Feeding	75	11.2	17	18.4	15	14.4	10	10.3	13	11.8	8	9.3	5	5.4	7	7.9
Other	219	32.6	24	26	35	33.6	29	29.9	23	20.9	25	29.1	49	52.6	34	38.6
Totals	672	100.1	929	9.4	104	99.8	9	100	110	100	86	100.1	93	100	88	99.6
Range	0–32.6%		0–26%		0–33.6%		0–29.9%		0–27.3%		0–29.1%		0–52.6%		0–38.6%	

Note: AM = Adult Male, AF = Adult Female, SAF = Sub-Adult Female, PAF = Pregnant Adult Female, and Juv = Juvenile (unsexed).

Table 2 State Data (Duration and Duration Percent) in minutes and seconds

Behaviour	Total	%	AM	%	AF	%	SAF	%	PAF	%	SAF	%	Juv	%	Juv	%
Grooming	25.59	23.6	1.3	5.1	3.8	19.9	8.3	21.2	8.9	52.2	2.8	25.2	.29	5.5	.20	3.4
Resting	16.29	15.0	6.5	25.7	4.3	22.5	3.4	21.4	4.0	13.2	2.5	22.5	.25	4.7	.10	1.7
Gazing	16.6	15.0	5	19.8	3.4	17.8	1.23	7.7	6.0	19.7	.45	4.1	.15	2.8	.34	5.9
Feeding	44	40.6	12.5	49.8	7.5	39.3	3.0	18.9	11.5	38.3	5.3	4 7.7	1.51	28.5	2.59	44.6
OS	5.8	5.3	——		.10	.5	——		——		——		3.15	59.4	2.57	44.6
Totals	108.3		23.3		19.1		15.9		30.4		11.1		5.3		5.8	
Range	5.3–40.6%		0–49.8%		0–39.3%		0–21.4%		0–52.5%		0–47.7%		0–59.4%		0–44.6%	

Note: AM = Adult Male, AF = Adult Female, SAF = Sub-Adult Female, PAF = Pregnant Adult Female, and Juv = Juvenile (unsexed). Results have been averaged over the two days observations.

DISCUSSION AND CONCLUSIONS

This study tends to support the findings of Hall and Mayer (1967). The adult male was less interactive with other members of the troop. This likely reflects his role as a guard against predators. Data are somewhat more inclusive in determining a female leader. Aggression was high between females 3 and 4 yet female 3 groomed the adult male more. Hall and Mayer (1967) found that the dominant female was in greater proximity to the adult male and groomed him more. However, they also found the leader female to be more aggressive. In the current study female #4 was the most aggressive. It may be that a power struggle was under way between females 3 and 4, or it may be that the pregnant condition of female 4 mad her more irritable. Yet another explanation is that due to competition for limited resources (in the wild), a pregnant female becomes aggressed against. Further study could clear this confusion since if her offspring gain a higher status than others she would probably have conferred this status onto the offspring. In addition, further study would determine if her behaviour changes when not pregnant.

Due to the limited scope of this study, conclusions must be drawn with caution, but it appears that female 3 or 4 is a leader over the other females and juveniles.

Note: This study did not investigate sounds related to aggression since they cannot be heard when the primates are inside.

References

Hall, K. R. L., & Goswell, M. J. (1964). Aspects of social learning in captive patas monkeys. *Primates*, *5*, 59–70.

Hall, K. R. L., & Mayer, B. (1967). Social interactions in a group of captive patas monkeys *Erythrocebus patas*. *Folia Primatologica*, *5*, 213–236.

Bibliography and References Cited

Alcock, J. (1989). *Animal behavior: An evolutionary approach* (4th ed.). Sunderland , MA: Sinauer Associates.

Alexander, R. D. (1975). The search for a general theory of behavior. *Behavioral Science, 20*(2), 77–100.

Altmann, J. (1974). Observational study of behavior: Sampling methods. *Behaviour, 49*, 227–267.

Altmann, S. A. (1968). Sociobiology of rhesus monkeys III: The basic communication network. *Behaviour, 32*(1–3), 19–49.

Anderson, D. R., Laake, J. L., Crain, B. R., & Burnham, K. P. (1979). Guidelines for line transect sampling of biological populations. *Journal of Wildlife Management, 43*, 70–78.

Asquith, P. (1986). Anthropomorphism and the Japanese and Western traditions in primatology. In J. Else & P. Lee (Eds.), *Primate ontogeny, cognition and social behaviour*. Proceedings of the 10th Congress of the International Primatological Society 3, 61–71.

Asquith, P. (1997). Why anthropomorphism is not metaphor: Crossing concepts and cultures in animal behavior studies. In R. W. Mitchell, N. S. Thompson, & H. Lyn Miles (Eds.), *Anthropomorphism, anecdotes, and animals* (pp. 21–34). Albany: State University of New York Press.

Bakeman, R., & Gottman, J. M. (1986). *Observing interaction: An introduction to sequential analysis.* Cambridge, U.K.: Cambridge University Press.

Bakeman, R., & Gottman, J. M. (1997). *Observing interaction: An introduction to sequential analysis* (2d ed.). Cambridge, U.K.: Cambridge University Press.

Bakeman, R., & Quera, V. (1995). *Analyzing interaction: Sequential analysis with SDIS and GSEQ.* Cambridge, U.K.: Cambridge University Press.

Barnett, A. (1995). *Expedition field techniques: Primates.* London: Expedition Advisory Centre, Royal Geographical Society.

Baulu, J., & Redmond, D. E., Jr. (1978). Some sampling considerations in the quantitation of monkey behavior under field and captive conditions. *Primates, 19*(2), 391–400.

Beattie, G. W., & Beattie, C. A. (1981). Postural congruence in a naturalistic setting. *Semiotica, 35*(1/2), 41–55.

Bernstein, I. S. (1991). An empirical comparison of focal and ad libitum scoring with commentary on instantaneous scans, all occurrence and one-zero techniques. *Animal Behavior, 42*, 721–728.

Box, H. (1984). *Primate behaviour and social ecology.* London: Chapman and Hall.

Boyd, H. L. (1998). *Postural congruence in a captive group of tonkean macaques (Macaca tonkeana).* Master's thesis, University of Calgary, Alberta, Canada.

Brim, J. A., & Spain, D. H. (1974). *Research design in anthropology: Paradigms and pragmatics in the testing of hypotheses.* New York: Holt, Rinehart, Winston.

Brockelman, W. Y., & Ali, R. (1987). Methods of surveying and sampling forest primate populations. In C. W. Marsh & R. A. Mittermeier (Eds.), *Primate conservation in the tropical rain forest* (pp. 23–62). New York: Alou R. Liss.

Burghardt, G. M. (1997). Amending Tinbergen: A fifth aim for ethology. In R. W. Mitchell, N. S. Thompson, & H. L. Miles (Eds.), *Anthropomorphism, anecdotes, and animals* (pp. 254–276). Albany: State University of New York Press.

Carpenter, C. R. (1934). A field study of the behavior and social relations of howling monkeys (*Alouatta palliata*). *Comparative Psychology Monographs, 10*, 1–168.

Carpenter, C. R. (1940). A field study in Siam of the behavior and social relations of the gibbon (*Hylobates lar*). *Comparative Psychology Monographs, 16*, 1–212.

Chapman, C. A., Wrangham, R., & Chapman, L. J. (1994). Indices of habitat-wide fruit abundance in tropical forests. *Biotropica, 26*(2), 160–171.

Coelho, A. M., Jr. & Bramblett, C. A. (1990). Behaviour of the genus *Papio*: Ethogram, taxonomy, methods and comparative measures. In P. K. Seth and S. Seth (Eds.), *Perspectives in primate biology* (Vol. 3, pp. 117–140). New Delhi, India: Today and Tomorrow's Printers and Publishers.

Colwell, R. K. (1974). Predictability, constancy, and contingency of periodic phenomena. *Ecology, 55*, 1148–1153.

Cords, M. (1993). On operationally defining reconciliation. *American Journal of Primatology, 29*(4), 255–268.

Darwin, C. (1872). *The expression of the emotions in man and animals.* Chicago: University of Chicago Press. (Reprint 1965)

Defler, T. R. (1995). The Time Budget of a group of wild woolly monkeys (*Lagothris lagotricha*). *International Journal of Primatology, 16*(1), 107–120.

de Lampe, M. G., Bergeron, Y. , McNeil, R., & Leduc, A. (1992). Seasonal flowering and fruiting patterns in tropical semi-arid vegetation of Northeastern Venezuela. *Biotropica, 24*(1), 64–72.

Delgado, R. R., & Delgado, J. M. R. (1962). An objective approach to measurement of behavior. *Philosophy of Science, 29*, 253–268.

DeVore, I., & Washburn, S. L. (1963). Baboon ecology and human evolution. In F. C. Howell & F. Bourliere (Eds.), *African ecology and human evolution* (pp. 335–367). Chicago: Aldine/Atherton Publishers.

de Waal, F. B. M., & van Roosmalen, A. (1979). Reconciliation and consolation among chimpanzees. *Behavioral Ecology and Sociobiology, 5*, 55–66.

Di Fiore, A., & Rendall, D. (1994). Evolution of social organization: A reappraisal for primates by using phylogenetic methods. *Proceedings of the National Academy of Sciences of the USA, 91*, 9941–9945.

Dyson-Hudson, R., & Dyson-Hudson, N. (1986). Computers for anthropological fieldwork. *Current Anthropology, 25*(5), 530–531.

Forney, K. A., Leete, A. J., & Lindburg, D. G. (1991). A bar code scoring system for behavioral research. *American Journal of Primatology, 23*, 127–135.

Frankie, G. W., Baker, H. G., & Opler, P. A. (1974a). Comparative phenological studies of trees in tropical wet and dry forests in the lowlands of Costa Rica. *Journal of Ecology, 62*, 881–913.

Frankie, G. W., Baker, H. G., & Opler, P. A. (1974b). Tropical plant phenology: Applications for studies in community ecology. In H. Leith (Ed.), *Phenology and seasonality modelling* (pp. 287–296). Berlin: Springer Verlag.

Hailman, J. P., & Strier, K. B. (1997). *Planning, proposing, and presenting science effectively.* Cambridge, U.K.: Cambridge University Press.

Hinde, R. A. (1973). On the design of check sheets. *Primates, 14*(4), 393–406.

Hinde, R. A. (1982). *Ethology: Its nature and relations with other sciences.* Oxford, U.K.: Oxford University Press.

Jazrawi, S. E. (2000). Postural congruence in a captive group of chimpanzees (*Pan troglodytes*). *American Journal of Primatology, 51* (Supplement 1): 25 (Abstract).

Johnson, S. M., & Bolstad, O. D. (1973). Methodological issues in naturalistic observation: Some problems and solutions for field research. In L. A. Hamerlynch, L. C. Handy, & E. J. Mash (Eds.), *Behavior change: Methodology, concepts, and practice* (pp. 7–67). Champaign, IL: Research Press.

Kappeler, P. M., & van Schaik, C. P. (1992). Methodological and evolutionary aspects of reconciliation among primates. *Ethology, 92*, 51–69.

Kohler, W. (1925). *The mentality of apes.* New York: Viking. (Reprint 1959)

Kraemer, H. C. (1979). One-zero sampling in the study of primate behaviour. *Primates, 20*, 237–244.

Krebs, C. J. (1999). *Ecological methodology* (2d ed.). Menlo Park, CA: Addison Wesley Longman.

Krebs, H. A. (1975). The August Krogh principle: "For many problems there is an animal on which it can be most conveniently studied." *Journal of Experimental Zoology, 194*, 221–226.

Kurup, G. U., and Kumar, A. (1993). Time budget and activity patterns of the lion-tailed macaque. *International Journal of Primatology, 14*(1), 27–40.

Lehner, P. N. (1979). *Handbook of ethological methods.* New York: Garland STPM Press.

Lehner, P. N. (1996). *Handbook of ethological methods* (2d ed.). Cambridge, U.K.: Cambridge University Press.

Lifshitz, K., O'Keeffe, R. T., Lee, K. L., & Avery, J. (1985). KIBOS: A microcomputerized system for the continuous collection and analysis of behavioral data. *Applied Animal Behaviour Science, 13*, 205–218.

Loy, J. D., & Peters, C. B. (1991). *Understanding behavior: What primate studies tell us about human behavior.* Oxford, U.K.: Oxford University Press.

Martin, P., & Bateson, P. (1986). *Measuring behaviour: An introductory guide.* Cambridge, U.K.: Cambridge University Press.

Martin, P., & Bateson, P. (1993). *Measuring behaviour: An introductory guide* (2d ed.). Cambridge, U.K.: Cambridge University Press.

Menon, S., & Poirier, F. E. (1996). Lion-tailed macaques in a disturbed forest fragment: Activity patterns and time budget. *International Journal of Primatology, 17*(6), 969–986.

Mitchell, R. W., Thompson, N. S., & Miles, H. L. (Eds.). (1997). *Anthropomorphism, anecdotes, and animals.* Albany: State University of New York Press.

Morisita, M. (1959). Measuring of interspecific association and similarity between communities. *Memoirs of the Faculty of Science Kyushu University Series E, 3*, 65–80.

Morisita, M. (1962). Id-index, a measure of dispersion of individuals. *Researches in Population Ecology, 4*, 1–7.
Napier, J. R., & Napier, P. H. (1967). *A handbook of the living primates.* London: Academic Press.
National Research Council (NRC) (1981). *Techniques for the study of primate population ecology.* Washington, DC: National Academy Press.
Odum, E. P., & Kuenzler, E. J. (1955). Measurement of territory and home range size in birds. *Auk, 72*, 128–137.
Overdorff, D. J., & Strait, S. G. (1996, August). Do prosimian primates function as seed dispersers in Madagascar? Paper presented at XVIth Congress of the International Primatological Society. Madison, WI.
Oxnard, C. E. (1983). *The order of man.* Hong Kong: Yale University Press and Hong Kong University Press.
Paterson, J. D. (1988). Computer based data collection. *Canadian Review of Physical Anthropology, 6*(2), 55–62.
Paterson, J. D. (1992). An alternative view: Behaviour as a multi-causal strategy for survival. In F. D. Burton (Ed.), *Social processes and mental abilities in non-human primates: Evidences from longitudinal field studies* (pp. 129–182). Lewiston, NY: Edwin Mellen Press.
Paterson, J. D. (1997). Seed predation patterns in the diet of the Sonso Forest baboon troop. Presented at annual meetings of the American Society of Primatologists, San Diego, CA, (June). Abstract: *American Journal of Primatology, 42*(2), 140.
Paterson, J. D., Kubicek, P., & Tillekeratne, S. (1994). Computer data recording and DATAC 6, a BASIC program for continuous and interval sampling studies. *International Journal of Primatology, 15*(2), 303–315.
Pileou, E. C. (1975). *Ecological diversity.* New York: John Wiley & Sons.
Rendall, D., & di Fiore, A. (1995). The road less traveled: Phylogenetic perspectives in primatology. *Evolutionary Anthropology, 4*(2): 43–51.
Renkonen , O. (1938). Statisch-okologische Unterschungen uber die terrestiche kaferwelt der finnischen bruchmoore. *Ann. Zool. Soc. Bot. Fenn. Vanamo, 6*, 1–231.
Rhine, R. J., & Ender, P. B. (1983). Comparability of methods used in the sampling of primate behavior. *American Journal of Primatology, 5*, 1–15.
Rhine, R. J., & Flanigan, M. (1978). An empirical comparison of one-zero, focal animal and instantaneous methods of sampling spontaneous primate social behavior. *Primates, 19*, 353–361.
Rhine, R. J., & Linvillr, A. K. (1980). Properties of one-zero scores in observational studies of primate social behavior: The effect of assumption on empirical analyses. *Primates, 21*, 111–122.
Rhine, R. J., Norton, G. W., Wynn, G. M., & Wynn, R. D. (1985). Weaning of free-ranging infant baboons (*Papio cynocephalus*) as indicated by one-zero and instantaneous sampling of feeding. *International Journal of Primatology, 6*, 491–499.
Sackett, G. P., Ruppenthal, G. C., & Gluck, J. (1978). Introduction: An overview of methodological and statistical problems in observational research. In G. P. Sackett (Ed.), *Observing behavior: Vol. II, data collection and analysis methods* (pp. 1–14). Baltimore: University Park Press.
Scheflen, A. E. (1964). The significance of posture in communication systems. *Psychiatry, 27*, 316–331.
Scheflen, A. E. (1965). Quasi-courtship behavior in psychotherapy. *Psychiatry, 28*, 245–257.

Scheflen, A. E. (1972). *Body language and social order: Communication as behavioral control.* Englewood Cliffs, NJ: Prentice Hall.

Scott, J. P. (1950). Methodology and techniques for the study of animal societies. *Annals of the New York Academy of Sciences, 51*(96), 1001–1122.

Silk, J. B. (1997). The function of peaceful post-conflict contacts among primates. *Primates, 38*(3), 265–280.

Sillen-Tullberg, B., & Møller, A. P. (1993). The relationship between concealed ovulation and mating systems in anthropoid primates: A phylogenetic analysis. *American Naturalist, 14*(1), 1–25.

Stearns, S. C. (1981). On measuring fluctuating environments: Predictability, constancy, and contingency. *Ecology, 62*(1), 185–199.

Stephenson, G. R., Smith, D. P. B., & Roberts, T. W. (1975). The SSR system: An open format event recording system with computerized transcription. *Behavior Research Methods and Instrumentation, 7*(6), 497–515.

Tinbergen, N. (1963). On aims and methods of ethology. *Zeitschrift für Tierpsychologie, 20*, 410–433.

Torgerson, L. (1977). Datamyte 900. *Behavior Research Methods and Instrumentation, 9*(5), 405–406.

van Hoof, J. A. R. A. M. (1967). The facial displays of the catarrhine monkeys and apes. In D. Morris (Ed.), *Primate ethology.* London: Weidenfeld and Nicolson.

Werner, D., Thuman, C., & Maxwell, J. (1992). *Where there is no doctor: A village health care handbook* (2d ed.). Palo Alto, CA: The Hesperian Foundation.

Wilson, C. C., & Wilson, W. L. (1975). Methods for censusing forest dwelling primates. In S. Konpo & A. Ehara (Eds.), *Contemporary primatology: Proceedings of the Fifth International Congress of Primatology* (pp. 345–350). Basel, Switzerland: S. Karger.

Wilson, E. O. (1975). *Sociobiology.* Cambridge, MA: Harvard University Press.

Wiseman, J. (1986/1996). *The SAS survival handbook: How to survive in the wild, in any climate, on land or at sea.* London: HarperCollins.

Wolfheim, J. H. (1983). *Primates of the world: Distribution, abundance, and conservation.* Seattle: University of Washington Press.

Wood, J. J. (1987). Some tools for the management and analysis of text: A pedagogic review. *Computer Assisted Anthropology News, 2*(4), 2–25.

Zuckerman, Sir S. (1932). *The social life of monkeys and apes.* Harcourt Brace, New York.

Resource References in Primatology

TITLE	Library of Congress CALL NO.
African Journal of Ecology	QL 750 E36
American Anthropologist	GN 1 A5
American Journal of Physical Anthropology	GN 1 A55
American Journal of Primatology	QL 737 P9 A125
American Naturalist	GN 1 A5
American Zoologist	QL 1 A52

Animal Behavior	QL 750 A5
Behavioral Ecology	QL 750 .B42
Behavioral Ecology & Sociobiology	QL 750 .B43
Behaviour	QL 750 BF
Biotropica	QH 1 .B54
Canadian Review of Physical Anthropology	GN 45 C34
Current Anthropology	GN 1 C85
Ecology	QH 540 .E3
Folia Primatologica	QL 737 P9 A1
International Journal of Primatology	QL 737 P9 A145
Journal of Animal Ecology	QL 750 J68
Journal of Anthropological Research	GN 1 S65
Journal of Comparative & Physical Psychology	BF 1 J58
Journal of Ecology	QH 540 .J6
Journal of Human Evolution	GN 281 A1 J68
Journal of Mammalogy	QL 700 J63
Journal of Medical Primatology	MED 13700
Journal of Theoretical Biology	QH 301 J75
Journal of Zoology (London)	QL 1 Z65
Laboratory Animal Science	MED 14690
Laboratory Animals	QL 55 A1 133
Laboratory Primate Newsletter	QL 737 P9 A15
Mammalia	QL 700 M34
Nature	Q1 N28
Primate Behavior	QL 785.5
Primate News	QL 737 P9 A163
Primates	QL 737 P9 A165
Primates in Medicine	QL 55 A1 P63
Quarterly Review of Biology	QH 301 Q3
South African Journal of Zoology	QL Z645
Yearbook of Anthropology	GN 1 Y42
Yearbook of Physical Anthropology	GN 60 Y4
Zeitschrift für Säugetierkunde	QL 700 Z43
Zeitschrift für Tierpsychologie	QL 750 Z43

About the Author

James D. Paterson and his wife, Sandra, have spent years in tropical forests, both in Uganda and Mexico. Dr. Paterson has studied captive, free-ranging Japanese Macaques (*Macaca fuscata*) at the South Texas Primate Observatory and the Oregon Regional Primate Research Center. He has also studied howler monkeys (*Alouatta palliata mexicana*) in Vera Cruz, Mexico. His primary interest remains the behavioral ecology of baboons (*Papio cynocephalus anubis*) in the Budongo Forest and surrounding high-grass savanna mosaic of Masindi District, Uganda. He has taught continuously (except for sabbaticals) at the University of Calgary, Calgary, Alberta, Canada, since 1971.

Two old observers, gumboots and all! At the Budongo Forest Project, Sonso clearing, Budongo Forest, Uganda, 1996. Courtesy of Rebecca Wingate-Saul and Lucy Bannon.